Atomic Physics for Everyone

Will Raven

Atomic Physics for Everyone

An Introduction to Atomic Physics, Quantum Mechanics, and Precision Spectroscopy with No College-Level Prerequisites

 Springer

Will Raven
Department of Physics
Smith College
Northampton, MA, USA

ISBN 978-3-031-69506-3   ISBN 978-3-031-69507-0   (eBook)
https://doi.org/10.1007/978-3-031-69507-0

This work was supported by Smith College

This Springer imprint is published by the registered company Springer Nature Switzerland AG
The registered company address is: Gewerbestrasse 11, 6330 Cham, Switzerland

If disposing of this product, please recycle the paper.

To the best person in the world
Prof. Maria-Teresa Herd

Acknowledgments

I have many people to thank for helping me write this book. To properly give praise, I need to provide a little background. I joined Smith College in 2013 and quickly became part of the AEMES (Achieving Excellence in Mathematics, Engineering, and Sciences) mentoring team, led by the amazing *Dr. Valerie Joseph*. This program is designed to increase diversity in STEM fields through a variety of initiatives. One of the key insights I gained was that providing students from underrepresented and/or under-resourced backgrounds with early research experiences is one of the best ways to keep them in STEM majors. Over time, the demand for research experiences grew among students. To address both of these needs—fulfilling the increasing demand for research experiences and providing early research opportunities to students from under-resourced backgrounds—I developed, with the encouragement and help of many people, a course-based research program. This program teaches first-year college students (freshmen) atomic physics, giving them time in the lab to conduct real science. The class has no college-level prerequisites; we use no calculus, linear algebra, or assume any background in introductory mechanics. However, we do use algebra and some trigonometry, specifically sines and cosines. The class was included in the Broader Impacts section of a National Science Foundation (NSF) grant that I was fortunate to be awarded: PHY-2110311.

People Who Helped Me Develop the Pedagogy for the Class I had casual conversations about class development with a large number of people, but there are a few who made substantial contributions including *Prof. Maria-Teresa Herd* (Physics, Assumption University), *Prof. Timothy Malacarne* (Sociology and Data Science, Nevada State University), and *Prof. Patricia DiBartolo* (Psychology and Director of the Sherrerd Center for Teaching & Learning, Smith College). The Smith College Office of the Provost and Dean of Faculty, led by *Prof. Michael Thurston* (English Language and Literature, Smith College), were incredibly supportive and provided personnel support to the physics department to free up my time to develop and teach the class.

My original intent was to write this book as free PDFs for my website. It was my NSF program manager *Dr. John D. Gillaspy* (Physics) who first encouraged me to publish the book for broader distribution. But, if my goal was to attract students from under-resourced backgrounds, it didn't make much sense to have them pay for a book. With additional support from Smith College, we are able to publish this book open access!

People Who Read the Book and Gave Me Lots of Feedback There are two people in particular who read the book multiple times through all the iterations. I owe *Prof. Maria-Teresa Herd* (Physics, Assumption University) and *Prof. Doreen Weinberger* (Physics, Smith College) so, so much. They were so insightful about approach and organization. This short paragraph seems incredibly insufficient for what they have done for me.

I want to thank the numerous people who read a selected chapter (or chapters). They are *Prof. Michael Thurston* (English Language and Literature, Smith College), *Prof. Gary Felder* (Physics, Smith College), and *Prof. Travis Norsen* (Physics, Smith College). And then there are all of the students who took this class with me. They found countless typos and mistakes, and pointed out confusing sections. More importantly, their questions and thoughtful exploration of the world of the super small helped me organize, expand, and enhance the book.

I also want to directly thank a few people who helped me in the classroom. These folks not only helped teach the class but also provided valuable feedback, sometimes in real time. Three of these individuals were learning assistants in the lab sections of the class the first time I taught it. They are *Bárbara Cabrales*, *Molly Herzog*, and *Chitose Maruko*. *Dr. Karl Ahrendsen* (Physics), who was a postdoctoral scholar in my lab, co-taught the class with me during the second iteration. He also gave an incredible amount of feedback to this text.

A Brief Tour of the Book

This book is divided into two parts. The first part, titled **Atom-Light Interactions**, explores spectroscopic techniques used to extract information about atoms. The second part, titled **Digging Deeper: Quantum Mechanics and Beyond**, examines the underlying physics of atomic structure, covering atomic notation, the principles behind it, current scientific understanding, and ongoing mysteries.

Generally, Part I is for learners interested in *how* we acquire knowledge about atoms, while Part II is aimed at those keen on understanding the fundamental physics of atoms. Readers may skip Part I if they are solely interested in the physics of atomic structure, but they should complete Chap. 1 before moving on to Part II.

The text also includes three appendices and a glossary that should serve as valuable references for readers. Appendix A is a periodic table, Appendix B is a long table that contains a lot of useful information about the atoms on the periodic table, and Appendix C lists all of the quantum mechanic rules that need to be satisfied for an electron to transition between two atomic states.

Contents

Part I   Atom-Light Interactions

1   Introduction to Atoms and Light 3
   1.1   What Is Atomic Physics? 4
   1.2   Conceptually Understanding the Atom 8
   1.3   Photons and Spectroscopy 17
   1.4   Math 19
   1.5   Extra: Polarization 20
   1.6   The Most Important Equation in All of Science 22
   Problems 22

2   "Natural Light" 25
   2.1   Breaking Light into a Spectrum 26
   2.2   Blackbody Radiation 30
   2.3   Discharge Lamps 35
   Problems 36

3   Atoms at Rest 41
   3.1   A Thought Experiment 42
   3.2   The Natural Linewidth in Angular Units 47
   3.3   The Natural Linewidth and the Lifetime of the Excited State 49
   3.4   The Scattering Rate and Saturation 50
   3.5   Power Broadening 56
   3.6   Example 57
   Problems 60
   References 62

4   Atoms in Motion 65
   4.1   The Doppler Effect 65
   4.2   Laser Frequency From an Atom's Perspective 67
   4.3   How the Velocity of an Atom Affects the Spectral Feature 72
   4.4   The Maxwell-Boltzmann Velocity Distribution 73
   4.5   Transmission and Absorption Plots for Atoms at a Non-Zero
          Temperature 76
   4.6   The Equipartition Theorem 77

    4.7   Application to Astronomy: Light from the Stars........................ 79
    Problems... 80

5   **Saturated Absorption Spectroscopy**.. 85
    5.1   Saturated Absorption Spectroscopy..................................... 85
    5.2   Crossovers... 90
         5.2.1 V Crossovers.. 91
         5.2.2 Λ Crossovers and X Crossovers.............................. 94
    5.3   Example with Cesium-133... 97
    5.4   Oxygen-16: A Spectrum Missing a Crossover........................ 100
    5.5   Example with Europium-151.. 101
    5.6   Extra: Crossover-Free Spectroscopy................................... 103
    Problems... 106
    References... 108

Part II   Digging Deeper: Quantum Mechanics and Beyond

6   **Quantum Mechanics vs. Classical Physics**.................................. 113
    6.1   What Is a State?.. 114
    6.2   Compatible vs. Incompatible Observables and the
         Uncertainty Principle.. 120
    6.3   Superposition of States.. 121
    6.4   The Energy Basis Set for a Quantum Harmonic Oscillator.......... 127
    6.5   The Uncertainty Principle Part 3....................................... 128
    6.6   Looking Forward... 130
    Problems... 131

7   **Angular Momentum**... 135
    7.1   Definitions.. 136
    7.2   Angular Momentum.. 138
    7.3   Orbital Angular Momentum of a Single Electron..................... 139
    7.4   The Magnitude and Projection of Angular Momentum.............. 144
    7.5   Adding Angular Momentum Vectors Together........................ 145
         7.5.1   Compatible or Incompatible?................................... 147
    7.6   Other Types of Angular Momentum.................................... 148
    7.7   A Bit More on Compatible and Incompatible Observables.......... 151
    Problems... 154

8   **Electronic Structure and Atomic Notation**................................. 157
    8.1   Energy Level Spacings... 158
    8.2   The Coulomb Interaction and Electron Shells......................... 160
    8.3   Term Symbols.. 165
    8.4   Connecting Angular Momentum to Orbitals........................... 170
    8.5   Fermions and Bosons.. 172
    Problems... 173

9   Hyperfine Structure 177
   9.1   Hyperfine Structure 177
   9.2   Math 181
   9.3   Transition Frequencies 184
   9.4   Example with Cesium-133 185
      9.4.1   Real Transitions 187
      9.4.2   Crossover Transitions 189
   9.5   Optional: Amplitudes 189
   9.6   Example with Oxygen-16 191
   9.7   Example with Oxygen-17 191
   Problems 192
   References 194

10   Isotope Shifts, Radioactive Decay, and the Nuclear Forces 197
   10.1   Isotope Shifts 198
   10.2   Radioactive Decay 202
   10.3   The Table of isotopes 207
   10.4   The Strong Nuclear Force 209
   10.5   The Weak Nuclear Force 212
   10.6   The Nuclear Shell Model and Energetics 216
   Problems 220
   References 223

11   The Standard Model of Particle Physics 225
   11.1   Problems with Quantum Mechanics 226
   11.2   The Uncertainty Principle Part 4 227
   11.3   Antimatter 229
   11.4   Going from Quantum Mechanics to Quantum Field Theory 231
      11.4.1   Remove the Conservation of Particle Number Constraint 231
      11.4.2   Vacuum Fluctuations 233
      11.4.3   Speed of Causality and Feynman Diagrams 234
      11.4.4   One More Thing 237
   11.5   The Standard Model of Particle Physics 237
   11.6   So, What's Next? 240
   Problems 242
   Reference 243

A   The Periodic Table 245

B   A Table of the Elements 247

C   Transition Rules: 251
Glossary 253

Index 261

Part I

Atom-Light Interactions

Introduction to Atoms and Light

1

Abstract

In this chapter, we explore the nature of light and atoms, focusing on their dual nature as both particles and waves. We examine why atoms have discrete energy levels and how only certain frequencies of light can excite electrons within these atoms. Through this exploration, we will understand the relationship between light's frequency, wavelength, and photon energy. We also explore key concepts such as wave interference and the historical experiments that shaped our understanding of quantum mechanics. Most importantly, this chapter emphasizes the scientific method, encouraging continual questioning of ideas, understanding, theories, and results to uncover the fundamental nature of the universe.

Learning Goals

By the end of this chapter, you should be able to understand:

- the basic structure of an atom.
- the concept of wave-particle duality and how it applies to both light and electrons.
- that atoms have discrete energy levels, and we can think about electrons as waves to understand why these energy levels are discrete.
- the historical experiments that led to the development of quantum mechanics, particularly the double-slit experiment.
- the basic principles of spectroscopy and its importance in studying atomic structure.
- that knowing the frequency of a laser, the wavelength of a laser, or the energy of a photon allows you to determine the other two. All three quantities are related by fundamental constants.

W. Raven, *Atomic Physics for Everyone*,
https://doi.org/10.1007/978-3-031-69507-0_1

1.1   What Is Atomic Physics?

Before we discuss **atomic physics**,[1] we should first ask a more fundamental question: What is **physics**? Before you read further, pause and think for a few minutes. If someone was to ask you, "What is physics?", how would you respond? My answer is below the word search:

Word Search!

E	S	H	T	L	E	A	P	Q	Q
I	C	U	I	H	A	H	Z	I	U
S	I	E	U	D	O	S	Z	W	A
P	S	P	L	T	D	N	E	O	N
E	Y	H	O	E	P	E	Z	R	T
C	H	N	A	L	C	U	N	D	U
T	P	T	U	M	O	T	A	S	M
R	K	S	U	O	U	R	R	G	X
U	G	N	M	W	K	O	K	O	E
M	P	R	O	T	O	N	U	Z	N

Word list:

ATOM

LASER

ELECTRON

PROTON

NEUTRON

QUANTUM

PHOTON

PHYSICS

SPECTRUM

Plus two hidden words

My answer: Physics is a branch of science that tries to understand the universe. We do this through exploring physical concepts and often explain those concepts using the language of mathematics.

All subfields of physics focus on a particular portion of the universe. For example, plasma physicists try to understand ionized gases, which are called plasmas. Atomic physicists[2] try to understand the world of atoms, which I like to refer to as the world of the super small.

Below are a few reminders that are important starting points for this book. If you aren't familiar with one of them, take 5 minutes to read the Wikipedia page on the topics.

---

[1] Words in the glossary are in bold type the first time we use them.

[2] In the last half of the twentieth century, atomic physics combined with molecular physics and optical physics to make what is now known as atomic, molecular, and optical physics. We use the acronym AMO physics.

- Atoms are composed of three types of particles: **protons**, **neutrons**, and **electrons**.
  - o Protons and neutrons form the center of the atom, called the nucleus, and electrons orbit around the nucleus.
  - o If an atom has an equal number of electrons and protons, the atom has no net charge. We call this a **neutral atom**. If we take an electron away from the atom or give the atom an extra electron, it is now charged. We call this an ion.
- **Molecules** are made from multiple atoms bound together. They can be the same atoms, like a nitrogen molecule consists of two nitrogen atoms, or different atoms, like a water molecule that has two hydrogen atoms and one oxygen atom.
- Atoms can be in the gas phase, liquid phase, or a solid phase.[3,4] Most atomic physicists study atoms in the gas phase to avoid the complexity of liquids and solids. In liquids and solids, atoms are bonded to one another, which makes the system more complicated. Our ideal system is a single atom all by itself away from all outside interactions.
- The periodic table of elements is a common way to view all the elements we know about. A copy of the periodic table can be found in Appendix A.
- Spectroscopists use light to interact with and learn about the world of the super small.

This book is aims to teach **quantum mechanics**, atomic physics, and **spectroscopy** without any advanced math. The most complicated math we use is algebra and some trigonometry (sines and cosines). Spectroscopy is a subfield of atomic physics that tries to learn about an atom through its interaction with light. Spectroscopy can be thought of as both a subfield and a tool. As a subfield, it is using atom-light interactions to try and understand the world of the super small. Additionally, it serves as a tool. The goal of atomic physics is to understand how atoms work, and spectroscopy is just a tool to accomplish this goal. The important thing is that spectroscopists use light to interact with atoms with the end goal of understanding the world of the super small. Scientists in lots of fields use spectroscopy, including atomic physicists, nuclear physicists, chemists, geologists, atmospheric scientists, and astronomers.

Atomic physicists try to reach this goal by starting with a simple system and building up complexity over time. One of our simplest systems is a single electron

---

[3] There are other phases of matter as well, including plasmas and Bose-Einstein condensates (BEC), which were named after the Indian mathematician and physicist Satyendra Nath Bose and the German born physicist Albert Einstein. A BEC is a really amazing state of matter that was first postulated in 1924. It wasn't until 1995 when American physicists Carl Wieman and Eric Cornell used atomic physics techniques to create the world's first BEC from rubidium atoms. Shortly after, German physicist Wolfgang Ketterle made a BEC from sodium atoms. The three atomic physicists won the 2001 Nobel Prize in Physics for this effort.

[4] One of my readers suggested that adding nationalities would add a bit of fun and extra history (I agreed!). Lots of sources list Einstein as 'German born' because he moved to Switzerland in 1895, giving up his German citizenship. In 1901, he became a Swiss citizen.

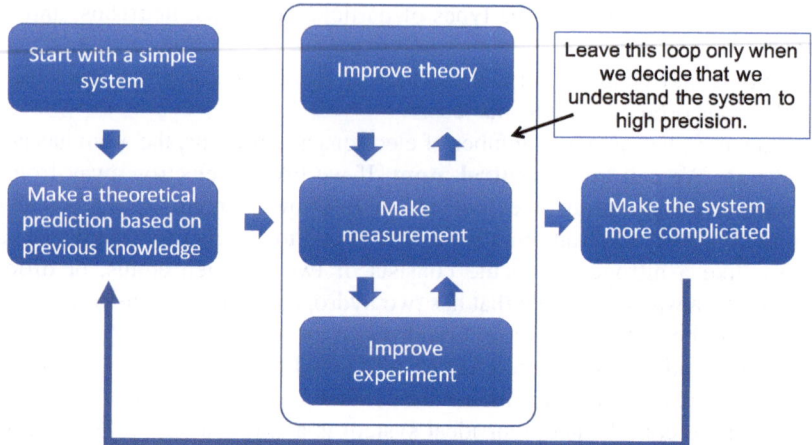

Fig. 1.1 A flowchart for trying to understand something. This is a version of the scientific method showing the interplay between experiment and theory

orbiting a single proton, also known as a hydrogen atom. More complicated systems include helium (2 electrons and 2 protons), lithium (3 electrons and 3 protons), neodymium (60 electrons and 60 protons), europium (63 electrons and 63 protons), and californium (98 electrons and 98 protons). Other fields of physics start with complicated systems and either "build up" or "build down" in complexity. For example, condensed matter physics, which includes subfields like superconductivity, has incredibly complex systems!

Figure 1.1 is a flowchart for how I like to explain how we, as scientists, try to understand something. For atomic physicists, our goal is to try to understand the world of the super small. We start with a simple system, for example the hydrogen atom. Next, we use theory that is based on previous knowledge to make a prediction. If the theory is good, the prediction should be confirmed by experiment. Once experimentalists make their first measurement, they start to improve the experimental setup to measure things better and better. While the experimentalists are improving their setup, the theorists are also improving their models. This process continues over and over and over until we, as a community, conclude that (1) the model is doing a really good job predicting the simple system, and (2) we run out of ways to make the experiment or theory more precise. If, after all that time stuck in that loop, we find that experiment and theory agree, we give out high fives and conclude we understand this simple system. If theory and experiment start to deviate from each other, physicists start to get excited because there is something we don't understand.

Once the atomic physics community is happy with the simple system, we make the system slightly more complex. For example, after we conclude that we understand hydrogen (1 electron, 1 proton, 0 neutrons), we move on to trying to understand deuterium (1 electron, 1 proton, 1 neutron) or helium (2 electrons,

2 protons, 2 neutrons). The most important thing to emphasize here is that if the model/theory is good, it should predict the experimental results before the experimentalists go and measure. If the theorists and experimentalists disagree, either the theorists messed up the math, the experimentalists messed up their experiment, or the theory is simply incomplete or all together wrong. The most exciting time in physics is when this last one happens.

The theoretical framework in atomic physics is quantum mechanics. Quantum mechanics has a more modern and complete version called the **Standard Model of Particle Physics**, the topic of Chap. 11. The Standard Model came about because, while quantum mechanics did a great job describing the world of the super small, it didn't do a perfect job. In other words, quantum mechanics isn't complete (it doesn't describe everything). As we worked through the flowchart, theory and experiment started to disagree as both theorists and experimentalists improved their methods. The Standard Model, which is far more mathematically complicated than Quantum Mechanics, does a much better job, but it is also not complete. For example, the Standard Model doesn't know how to describe lots of things we observe in real life like gravity, dark matter, dark energy, and baryon asymmetry, all of which will be discussed in Chap. 11. Despite these problems, experimental measurements that test the validity of the Standard Model always seem to confirm the Model is accurately describing nature![5] Therefore, we test it over and over again hoping that (1) we find a disagreement between experiment and theory and (2) this discrepancy leads to a more complete model describing the world of the super small.

In this book, we are going to think about atoms in their gaseous form, so we don't have to worry about how two or more atoms interact in a molecule, a liquid, or a solid. If you are reading this book as part of a class, you may also have an experimental portion where you are focusing on a single atom like sodium, cesium, europium, or neodymium. One of the most important things to note right now is that we want to know general properties of the atom of choice. For example, if we want to know the energy separation between two atomic states (more on this below) of a single, isolated atom, we want to know that energy separation in the absence of external interactions. We don't care about these properties in a magnetic field, an electric field, or even a laser field. Imagine a single atom in the middle of outer space, completely in the dark, and far from any other object. We want an experiment that helps us figure out these general properties. If a scientist knows the general properties, they can then calculate or estimate what would happen to that property if the atom was, for example, put in a magnetic field. The important thing is that we provide the general information that can later be used by theorists, other experimentalists, and engineers.

---

[5] To be fair, experiment does not always agree with the Standard Model. For example, the Standard Model says that the universe should be nearly equal amounts of matter and antimatter (this is the baryon asymmetry listed above), which would be bad since then all the matter and antimatter would combine to destroy the universe. In other words, the Standard Model says that the universe shouldn't exist, which disagrees with experiment.

1.2   Conceptually Understanding the Atom

We are not going to discuss the Standard Model right now. We are going to keep things a bit simpler and discuss, conceptually, the atom. Later on in the book, we will add to our conceptual model to make things more complete. So, what does quantum mechanics say about the atom?

To answer that question, let's first think about an electron that is orbiting around a nucleus, as shown in Fig. 1.2. The electron has negative charge and the nucleus has positive charge. According to electromagnetic theory, the electron should radiate away energy. Imagine if you were flying a hand glider and you slowly lost energy. In this scenario you would slowly drift down until you landed on the ground. According to electromagnetic theory, the electron will also lose energy and should orbit closer and closer to the nucleus until it collides with the nucleus. Experiment shows this is not true. In fact, if it were true, we wouldn't be here reading this book because atoms wouldn't exist and thus the universe wouldn't exist,[6] which disagrees with experimental observation.

There were some amazing experiments conducted in the late 1800s and early 1900s that seemed to imply that the electron is *not* like the moon orbiting around the earth. One of my favorite experiments that showed this behavior is called the double-slit experiment. If we assume electrons are like little sticky balls and send them through two small slits in a barrier, as shown in Fig. 1.3, we expect to have two strips of balls stuck to the screen. However, that isn't what is seen experimentally!

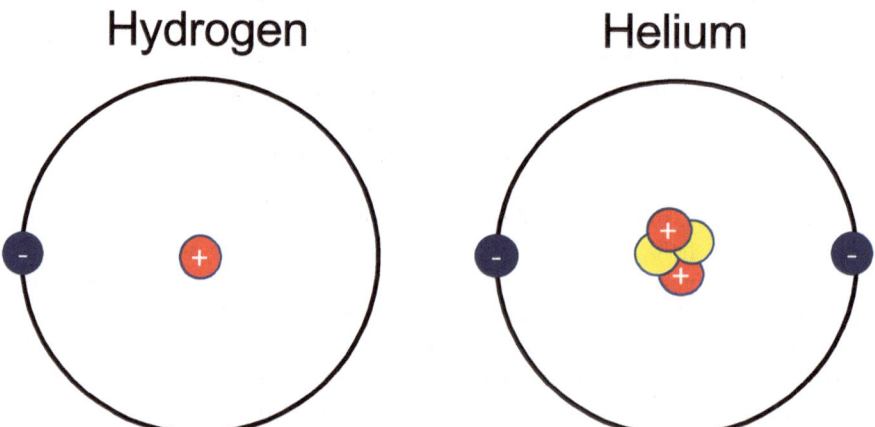

Fig. 1.2 A helpful conceptual, but incorrect, way of thinking about atoms. The model where the electrons orbiting around the nucleus is called the Bohr model, named after Danish physicist Niels Bohr

---

[6] Well, maybe the universe would exist. It would just exist without any matter, which wouldn't be much fun.

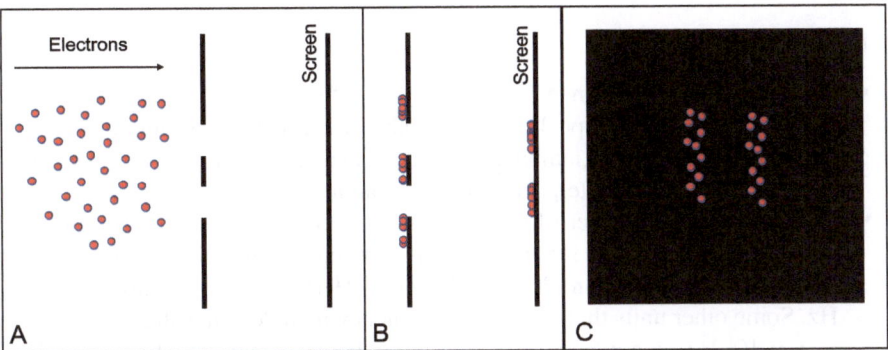

Fig. 1.3 A double-slit experiment where the electrons behave like little sticky balls. In panel A, the electrons are moving towards a barrier that has two small slits in it. Behind the barrier with two small slits is a screen with no holes. In panel B, the electrons have crashed into or passed through the barrier to crash into the screen. Panel C shows a front view of the electrons that passed through the barrier and crashed into the screen

Before we get to the real experimental results, we need to define some important terms and explore the concept of **wave interference**. Below are three important definitions about waves. Figure 1.4 is a visual representation of two of those definitions: wavelength and frequency.

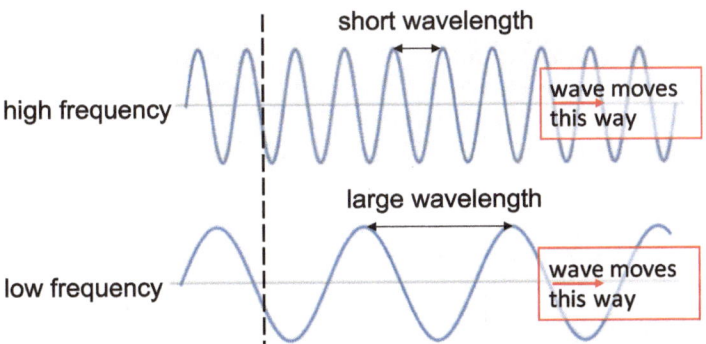

Fig. 1.4 A visual description of wavelength and frequency. Wavelength is the distance between two "like" points on the wave, for example the distance between two adjacent peaks or two adjacent troughs. If the wave is moving towards the right, frequency is how many peaks pass through the dashed line every second. If both waves are moving with the same speed, the upper wave has a higher frequency than the lower wave since more peaks pass the dashed line in 1 second

Definitions

- **Wavelength:** The distance between any two "like" points on a wave, such as two adjacent wave peaks. The variable we use for wavelength is the lowercase Greek letter lambda, λ. Since wavelength is a distance, we use length units such as meter, centimeter, or nanometer.
- **Frequency:** The number of oscillations per second. We use the variable f for frequency. The unit for frequency is 1/seconds, which is called a hertz, named after the German physicist Heinrich Hertz. A hertz is shortened to Hz. Some other units that we use for frequency include a megahertz (1 MHz $= 1 \times 10^6$ Hz), a gigahertz (1 GHz $= 1 \times 10^9$ Hz), and a terahertz (1 THz $= 1 \times 10^{12}$ Hz).
- **Period:** The time it takes for the wave to complete one full oscillation. We use the variable T for period, and the unit is seconds. Frequency and period are related by the formula $T = 1/f$.

Wave Interference Wave interference happens whenever two or more waves overlap. Those waves can be traveling in the same direction, opposite directions, or at an angle with each other. In fact, they don't even have to look like the waves in Fig. 1.4. They could instead be single pulses like in Fig. 1.5. In this example, the two pulses, represented by the blue and red dashed lines, are on the same rope (the thicker gray line) and traveling in opposite directions. When the pulses overlap, they constructively add to make a larger pulse. The is known as constructive interference.

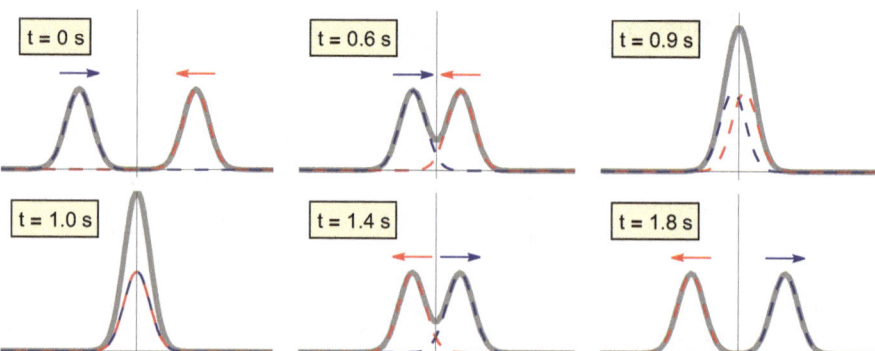

Fig. 1.5 Two pulses that constructively interfere with each other: The thick gray line is what we would actually see. The blue dashed line shows the pulse traveling to the right while the red dashed line shows the pulse traveling to the left. When they pass through one another, they add to create a larger pulse. At $t = 1.0$ s, this is 100% constructive interference. At all other times, the pulses are only partially constructively interfering

However, if we have a positive amplitude pulse and a negative amplitude pulse, we will get destructive interference, see Fig. 1.6.

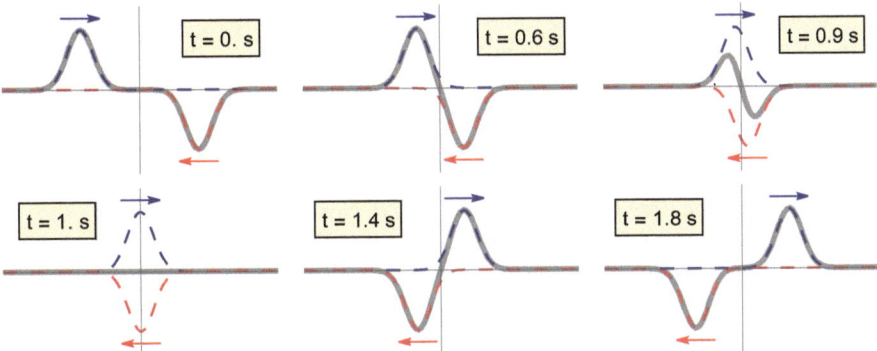

Fig. 1.6 Two pulses that destructively interfere with each other: The thick gray line is what we would actually see. The blue dashed line shows the pulse traveling to the right while the red dashed line shows the pulse traveling to the left. When they pass through one another, they perfectly cancel each other out creating no disturbance on the actual rope for a brief period of time. At $t = 1.0\,$s, this is 100% destructive interference. At all other times, the pulses are only partially destructively interfering

Try This

Find a piece of thin rope or a slinky and a stopwatch. If you have a friend nearby, have them hold one end of the rope or slinky. If not, tie or connect one end of the rope or slinky to a door knob. Stand a distance apart so that there is a bit of tension on the rope or slinky. Send a pulse down the rope or slinky and watch what happens when the pulse reflects off your friend or the doorknob. Next, start creating a sine wave motion with your hand, see Fig. 1.4. Try to move your arm up and down so that the rope or slinky creates the shapes seen in Fig. 1.7.

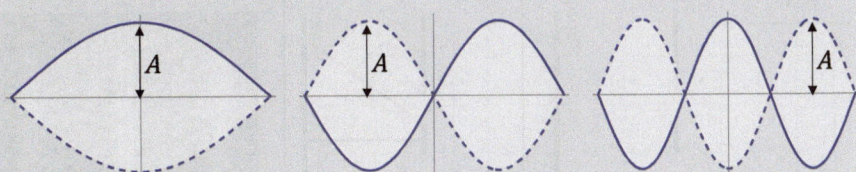

Fig. 1.7 The first three standing waves of a one dimensional rope or slinky. Each standing wave is drawn with the same amplitude A. Interestingly, the frequency of a standing wave is independent of the amplitude of the standing wave

(continued)

You will find that your hand has to move up and down with a very specific frequency to create these shapes, which are called standing waves. When you produce standing waves, the wave that you are creating with your hand is constructively interfering with the reflected wave. No other frequencies produce perfect constructive interference. Using your stopwatch, find the time for ten full oscillations (the time it takes for your hand to move up and down ten times) for each of the first three standing waves. The period of the standing wave is that time divided by 10. If you count twenty full oscillations, you would divide the time by 20 to find the period. Next use the formula $f = 1/T$ to find the frequency needed to produce that standing wave. The frequency of the standing wave with one "loop" (the left picture) is called the fundamental frequency. You should find that the second picture with two loops has twice the fundamental frequency. The third picture with three loops should have a frequency that is three times the fundamental frequency. In general, $f_n = nf_1$, where n is how many loops the standing wave has. Notice that the higher mode standing waves (the waves with more loops) require more shaking energy! This will be important later.

Back to the Double-Slit Experiment Ok, let's run a different experiment. Instead of little, sticky balls, let's send a wave towards the screens, as shown in Fig. 1.8. Panel A shows the wave traveling towards the slits. The vertical lines are supposed to indicate the peaks of the wave. The troughs are halfway between each peak, so you are visualizing the wavelength of the wave. Panel B shows the wave as it passes through the slits. The straight wave turns into two arc waves, one coming from each slit. The arc waves are basically half circles with the center of the circle at the slit. I changed the color of one of the circular waves to better visualize the evolution of each wave. The two circular waves interfere with each other on the screen. Places

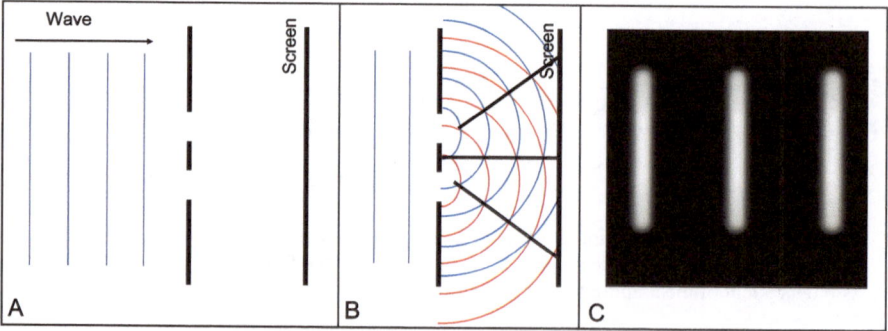

Fig. 1.8 A double-slit experiment using waves

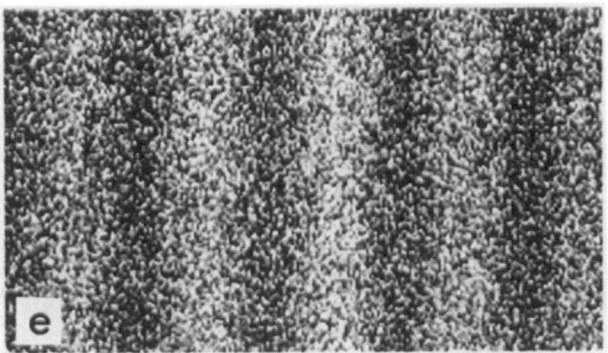

Fig. 1.9 Experimental results of a double-slit experiment with electrons. Image Credit: Dr. Tonomura and Belsazar from Wikimedia Commons CC BY-SA 3.0

where peaks of the waves overlap, indicated by the black lines between the slits and the screen, are where the two circular waves constructively interfere. Half way between the points of constructive interference are points of perfect destructive interference. Panel C shows what we see on the screen. The bright parts are where the two waves constructively interfere. The dark parts are where the two waves destructively interfere. At the very center of the dark parts, the waves are 100% destructively interfering.

Now, let's do the experiment with electrons! Real experimental results can be seen in Fig. 1.9. The electron behaves like a wave! There is actually a lot more to the double-slit experiment, but we aren't going to go into any more detail in this book. You will learn much more about it if you take a Modern Physics class. The really important thing we need to conceptually understand is that the electron has wavelike behavior. This is incredibly important information for us because an electron that is orbiting a nucleus is *not* like the moon orbiting the earth, which is depicted in Fig. 1.2. Experiments show that electrons behave like waves.

▶ **Important Concept** Electrons in atoms behave like waves, so we need to think about interference effects.

This is a really hard concept to wrap our brains around, but experiments seem to indicate this idea is correct. How does a wave "orbit" around a nucleus? As an analogy, imagine a wave that wraps around upon itself in a circle, see Fig. 1.10. This figure is just a conceptual example since there is no start or end to the wave. But to explore this concept we are going to wrap the electron wave counter-clockwise around the nucleus, and the wave is going to interfere with itself. If the electron wave does not perfectly wrap back around so that two peaks don't overlap perfectly, the electron will destructively interfere with itself. If an electron destructively interferes

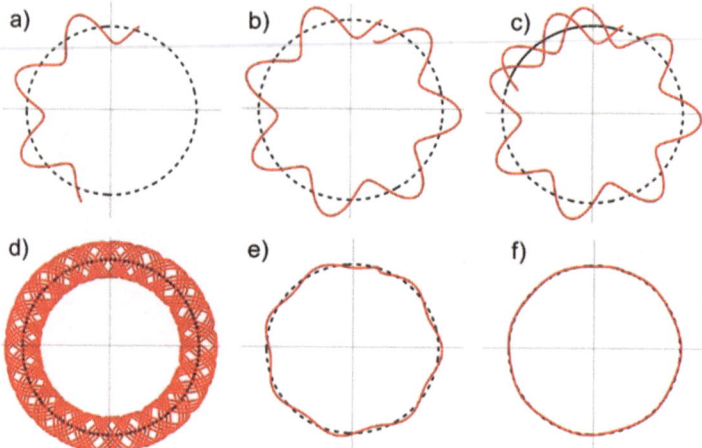

Fig. 1.10 A conceptual exploration of an electron wave orbiting a nucleus that would result in destructive interference. In (**a**), we imagine the electron wave starting at the top of the circle and traveling counter-clockwise. In (**b**), the electron wave has made one full orbit, but notice that the wave does not line up with its starting point. For (**c**), the wave continues and should be interfering with itself, but we aren't going to add the waves together quite yet. In (**d**), the wave continues for 12 orbits. To explore destructive interference, (**e**) shows adding together this wave after 4 full rotations. (**f**) shows destructive interference of this wave after 10 full rotations

with itself, there is no wave! If there is no wave, then there is no electron. Since electrons exist, the electron wave must constructively interfere with itself.

A lot has happened, so let's do a quick recap. We have learned that:

- Electrons have wavelike properties (Fig. 1.9)
- Waves interfere with one another (Figs. 1.5, 1.6, and 1.7)
- A wave can also interfere with itself (Figs. 1.8 and 1.10)
- An electron exists, so it better not destructively interfere itself out of existence.

Pause here and see if you can come up with a conclusion. When you have reached a conclusion, read on. My conclusion is below this fun fact:

Fun Fact
Many musical instruments make noise (hopefully pleasant noise!) because of standing waves. For example, when a guitarist plucks a guitar string, which is fixed at both ends, they are creating many standing waves (see Fig. 1.7) with different amplitudes all at the same time. Each of those standing waves has a frequency that is a multiple of the fundamental frequency. The amplitude of each of those standing waves is what gives the guitar it's distinctive sound. To

(continued)

play a different note, the guitarist changes the length of the string by pressing on the string in a different spot, which in turn changes the fundamental frequency.[a] The same is true for a piano. A piano key strikes a wire that produces standing waves that result in a (hopefully!) pleasant noise. The amplitudes of all the standing waves on a struck piano wire are different than the amplitudes of a guitar string, which is why they sound different.

A note played on a trumpet or saxophone also produces standing waves, but the standing waves are created in the air and are physically manifested as places of high and low air density.

---

[a]They could also pluck a different string that has a different mass density or change the tension on the string. All of these changes will result in a different fundamental frequency.

Conclusion The electron must constructively interfere with itself. This idea is conceptually shown in Fig. 1.11.

An electron can only "orbit" around the nucleus if it satisfies a standing wave condition. Each standing wave has an energy associated with it. Just like how the standing wave on a string with 2 loops has a shorter wavelength and requires more shaking energy than the standing wave with 1 loop, the higher energy "states" of an electron have shorter wavelengths. Thus the allowed energies of the electron are "discrete": they can only have the specific values corresponding to these standing waves.

We call the discrete energies an electron orbiting a nucleus can have "energy levels" or "energy states" and describe them in diagrams like Fig. 1.12. The lowest energy state is called the ground state of the atom, or the ground state for short. States with higher energy are called excited states. These energy level pictures are sometimes called Grotrian diagrams, named after German astronomer and astrophysicist Walter Grotrian. The SI unit for energy is a joule, which is named after

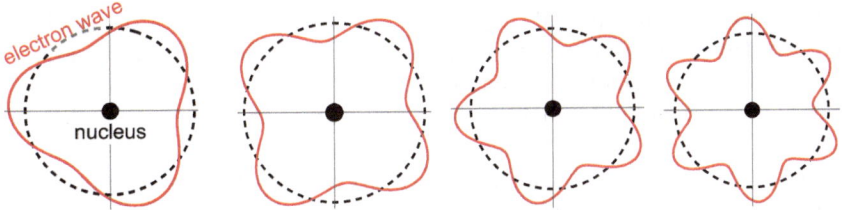

Fig. 1.11 Four conceptual examples of an electron wave orbiting a nucleus that would result in constructive interference. Going from left to right is going from a lower energy state to higher energy states

Fig. 1.12 A conceptual picture that shows an atom with a single ground state and three excited states. A real atom has many excited states

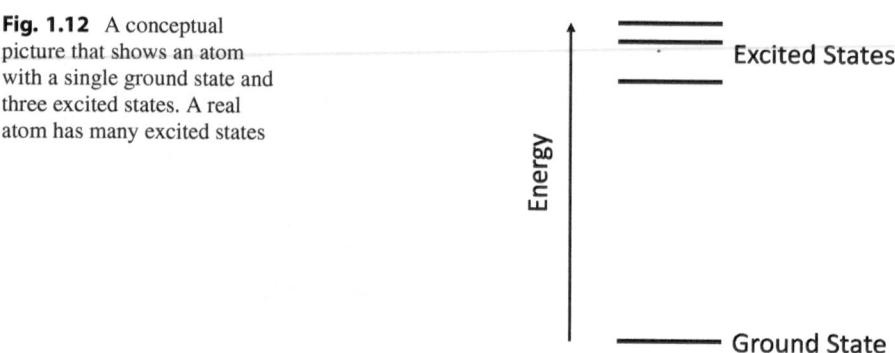

the English physicist James Joule. A joule is shortened to J.[7] A joule is shorthand for $kg\,m^2/s^2$. For reference, the energy of a baseball moving at 50 mph is about 36 joules. The energy of an apple moving at 1 meter/second is about 0.05 joules.

▶   **Important Comment**   Atomic physicists and spectroscopists often say that the ground state energy is 0. This is not true! Think about the standing wave on a string with a single loop. This is the standing wave with the smallest amount of shaking energy, but it still has energy! The ground state of an atom also has energy. In experiments, we measure the energy difference between the ground state and excited states. For this purpose, we assign the lowest energy state to be 0 so that everything is measured with respect to that state.

The energy difference between the ground state and the lowest energy excited state in the hydrogen atom is 0.000000000000000001634 J $= 1.634 \times 10^{-18}$ J. That is a small number! A more common unit for energy in atomic physics is the electronvolt (eV). The conversion is 1 eV $= 1.602 \times 10^{-19}$ J, so the energy of the lowest excited state is 10.2 eV above the energy of the ground state.

One Final Thing   The above description of an electron in an atom is a good starting point to understanding how and why atoms behave the way they do. Think about this chapter as making a first pass-through Fig. 1.1. The conceptual idea you just learned is correct. However, the electron wave turns out to be more complicated than the simple picture given above. For one thing, the electron wave is going to be in 3 dimensions. But, more importantly, the actual wave that describes the electron in an atom, which is called a wavefunction, is a bit messier to deal with. All of the concepts like wave interference and energy levels in the atom are still correct, but the presentation of the electron wave has been simplified. You will refine these ideas in future classes such as Modern Physics and Quantum Mechanics.

---

[7] Notice the unit 'joule' is not capitalized. In general, a unit named after a person is not capitalized.

A Little Bit Extra

A careful reader might have noticed something odd about Fig. 1.9. While there was clearly a wave interference pattern, this wave interference is composed of tiny bright dots. The electrons are moving through the slits as waves. However, we have experimentally found that when the electron hits the screen, which is often called a measurement, the electron wave "collapses" to a single point. You may have heard the phrases "wavefunction collapse" or "**wave-particle duality**". This is a fancy way of saying that the electron behaves like a wave until something interacts with it before it starts behaving like a particle (i.e. a single point on the screen). The amplitude of the wave interference tells us the *probability* of where the wave collapses back to a point. Thus where the electron wave destructively interferes on the screen produces 0 probability that the particle will be detected at that point. The process of going from a wave to a point is a fascinating and yet unexplained phenomenon! There are many physicists and philosophers who think very hard about how this "collapse" occurs. We will explore more about the probabilistic interpretation of quantum mechanics, which is often called the orthodox or Copenhagen interpretation, in Chap. 6.

1.3 Photons and Spectroscopy

Suppose we want to measure the energy difference between the ground state and any one of the excited states. How do we do it? Answer: Spectroscopy! We shine light onto the atoms and look at what happens to the light. In modern spectroscopy, we use laser light. Light, including laser light, is composed of tiny particles known as **photons**.[8] As an analogy, think about a stream of water. The water looks continuous, but it is actually made up of tiny water molecules. The same thing is true with light. Light is composed of tiny particles we call photons. There is, however, a really important difference between water molecules and photons. The energy of a photon is determined only by the frequency of the light wave. The more the light wave oscillates up and down in one second, the larger the energy of an individual photon that makes up that light wave. The energy of a water molecule, on the other hand, is related to how fast the molecule is moving, rotating, and vibrating. We will explore the connection between the frequency of the light wave and the energy of a photon more in the next section.

---

[8] German physicist Max Planck won the 1918 Nobel Prize in physics for discovery of the photon. Albert Einstein won the 1921 Nobel Prize in physics for explaining the photoelectric effect, which determined that a beam of light is made up of a bunch of photons. I may be biased, but Einstein's explanation of the photoelectric effect was the most important discovery of the twentieth century.

▶ **Important Statement** If a photon has the same energy as the energy
difference between the ground state and an excited state, the atom will
absorb that photon and move an electron to the excited state. If the
photon does not have the same energy as that energy difference, the
atom will completely ignore the photon. [9]

Thought Experiment Time Imagine you have an electron behaving like a wave
orbiting around a nucleus. That electron wave has to constructively interfere with
itself, and we can think about that standing wave as having some amount of energy.
Let's call this energy E_1. If we wanted to add another loop to the standing wave,
we need to add energy to the system. Suppose an electron with 1 more standing
wave loop has energy E_2. Remember that the electron needs to be a standing wave,
which has a specific energy. If it wasn't, the electron would experience destructive
interference. We want to move an electron from the lower energy standing wave
to the higher energy standing wave. How much energy do we need to put into the
system? The answer is in the footnotes.[10]

This is the basic idea behind spectroscopy. If we want to excite an electron from
one energy level, which has some energy, to another energy level, which has a
different energy, we need to provide the system with the correct difference in energy.

Restating that in terms of an actual experiment: We send a laser through an
optically clear container, called a cell or vapor cell, filled with atoms. We will
smoothly scan the energy of the photons over time and monitor the transmission
of the laser through the cell. If the energy of the photons does not match the energy
difference between the ground state and an excited state, the laser light will pass
right through the atoms with no losses. If the energy of the photons matches that
energy difference, light will be lost from the laser and the transmission will decrease.
Experiments that match the above description are known as absorption spectroscopy
experiments. We, as experimentalists, simply monitor the transmission of the laser
through an atomic sample as we change the photon energy. When the amount of
transmitted light drops, we learn what energy is required to excite the atoms from
the ground state to an excited state. Believe it or not, that is, conceptually, all there
is to it. In the lab things are a bit more complicated, but this is the basic idea. We
will finish up this section with a new definition.

---

[9] This important statement is super important, but it also isn't 100% correct. We are going to start
with this statement to get at some important concepts. In the next few chapters, we are going to
discuss some more physics and then restate this important statement to something more correct.
It's kind of like the flowchart from the beginning of the chapter. We start simple and build up
complexity.

[10] We would need to add precisely $E_2 - E_1$ of energy.

Definitions

- **Resonance:** When an atom gets excited by a photon from one state to another, we say the atom "goes through resonance." This is similar to playing the trumpet. When you blow correctly into a trumpet, a standing wave is excited in the pipe to create a note. The same thing happens with an atom. When you excite an atom with the right energy photon, the electron goes from one standing wave mode to another. Atomic physicists use the word "excitation" and "resonance" interchangeably.

1.4   Math

The speed of light, the wavelength of the laser light, and the frequency of the laser light are all related using the formula:

$$c = f\lambda, \tag{1.1}$$

where $c = 299,792,458 \, \text{m/s} \approx 3 \times 10^8 \, \text{m/s}$ is a constant of nature known as the speed of light.[11] The energy of a photon is given by the formula:

$$E_{\text{ph}} = hf, \tag{1.2}$$

where h is a constant of nature known as Planck's constant, named after German physicist Max Planck. Planck's constant is a very small number, $h = 6.626 \times 10^{-34} \, \text{Js}$. The units are joules times seconds, where joules is the unit of energy. The most important thing to emphasize here is that we have multiple ways to state the same property of a photon. If I tell you the wavelength of the light, you can immediately calculate the frequency of the light and the energy of the photons that make up that light. Since all three quantities are related by constants, an atomic physicist will claim that wavelength, frequency, and the energy of a photon are all the same thing. An atomic physicist might very well say, "the energy of the photon is 632 THz" and think nothing of it. It takes some practice to use "wrong" units to describe something! Imagine if you asked me my height, and I responded with 150 MHz. For physicists, there is nothing wrong with doing this since we would only be off by a constant of nature.

A common energy unit is inverse centimeters. For example, the statement, "The energy difference between the two states is $12,460 \, \text{cm}^{-1}$" is very common. For spectroscopists, this number means the same thing as $2.475 \times 10^{-19} \, \text{J}$. Converting

---

[11] This is the speed of light in a vacuum. It is exactly $299,792,458 \, \text{m/s}$.

between the two numbers requires multiplying by the correct combination of the fundamental constants h and c. This energy unit is explored more in Problem 1.6.

We can now make a new definition:

Definitions

- **Resonance frequency:** When the frequency of the laser is just right to excite the atom from the ground state to an excited state, we call that the resonance frequency. We use the variable f_r to represent resonance frequency. Since resonance frequency is a frequency, it uses units of Hz, MHz, GHz, or THz.

Common Misconception   Power is not the same thing as energy. Power is the *rate* at which energy is exiting the laser. Imagine two laser beams with two different frequencies. The light from laser #1 has a higher frequency than the light from laser #2. We now know that the photons that make up laser #1 have more energy than the photons that make up laser #2. Power is how much energy is exiting the laser per second. If the same number of photons per second are leaving both lasers, laser #1 has a larger power. In math form, the power of a laser is:

$$P = N E_{ph}, \tag{1.3}$$

where N is how many photons per second that leave the laser and $E_{ph} = hf$ is the energy of a single photon. The unit for power is a watt, named in honor of Scottish chemist James Watt. A watt is shortened to W and is equivalent to joules per second. Imagine you have a 1 watt laser and a 10 watt laser that have photons with the same energy. The 10 watt laser emits 10 times as many photons per second as a 1 watt laser.

1.5   Extra: Polarization

The frequency of light tells us how many times that light wave oscillates up and down in 1 second. Polarization tells us the direction the light is oscillating. There are three major groupings of light polarization: linear, circular, and elliptical. Linearly polarized light is the most common light we use in spectroscopy, so we aren't going to talk about circularly or elliptically polarized light here. However, there are types of spectroscopy experiments that do use circularly polarized light. If you are curious about them, image search the phrases to find some neat animated gifs showing light with different polarizations moving through space. For now, we will focus on linearly polarized light.

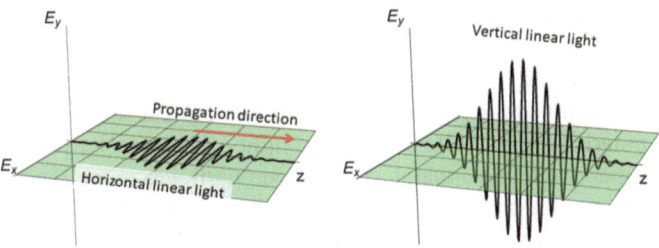

Fig. 1.13 Examples of oscillating electric fields propagating along the z-direction. The left picture shows horizontal linear light while the right picture shows vertical linear light

Linear polarization is light that is oscillating up and down in a single plane. There are two specific types of linearly polarized light: horizontal and vertical. Horizontal light oscillates . . . horizontally with respect to some surface, and vertical light oscillates . . . vertically to that surface, see Fig. 1.13 Remarkably, we can use these two polarizations to describe any linearly polarized light. For example, suppose the light was oscillating at a 45° angle. We would describe the light as half horizontal and half vertical.

In the lab, there are optical devices called half-waveplates, sometimes written as $\lambda/2$ plate or just $\lambda/2$. A half-waveplate can rotate the linear polarization of light. If you have horizontally polarized light, you can use a half-waveplate to change the polarization so that it is 10% vertical and 90% horizontal, 50% vertical and 50% horizontal, 75% vertical and 25% horizontal, etc. You can even make the light exiting the half-waveplate be completely vertical. That might seem like a neat trick, but the real usefulness comes when we put a second optical device after the halfwave plate called a polarizing beam splitter (PBS), see Fig. 1.14.

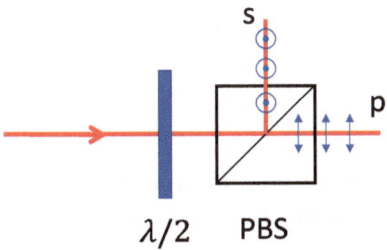

$\lambda/2$ PBS

Fig. 1.14 Using a half-waveplate and a polarizing beam splitter, we can create two beams of light. You can rotate the half-waveplate to control the ratio of light in each path. In normal spectroscopy setups, you are looking down on the light and optics from above. In this orientation, p-polarized light is horizontal, indicated by the blue arrows and s-polarized light is vertical, indicated by the dotted circles

By tradition, the light that bounces off the PBS is called s-polarized light while the light that passes through is called p-polarized light.[12] This is because the PBS can technically be in any spatial orientation. For safety reasons, we almost always keep the light in a horizontal plane. In this typical setup, s-polarized light is vertically polarized light while p-polarized light is horizontally polarized light. If you were to rotate the PBS so that the s-polarized light was going straight up (don't do this, it is an eye hazard!!), that s-polarized light is now horizontally polarized while the p-polarized light is now vertically polarized. This is why we use "s" and "p".

In the lab, we use half-waveplates and polarizing beam splitters to split a single laser beam into two beams and also control the power in each of them. If the waveplate is oriented such that the light polarization leaving the waveplate is completely vertical, all of the light would bounce off the PBS. If the polarization of the light leaving the waveplate was at 45°, there would be equal power in both lasers leaving the PBS.

1.6   The Most Important Equation in All of Science

questions + repetition + critical thinking = mastery

Don't be afraid to ask questions. Don't be worried about asking for clarification. Don't expect to remember or understand every single concept the first time you read or hear about it. Mastery is not a short journey. Developing critical thinking skills is not a 5 minute activity that you figure out after watching a 3 minute YouTube video. Practice deep learning, keep a growth mindset, and, most importantly, have fun!

Problems

1.1 Go through the chapter and write down all of the fundamental constants (there are two of them). Don't forget units.

1.2 Go through the chapter and write down each equation. For each equation, write a brief description about what the equation means.

---

[12] Fun fact: the "s" stands for the German word "senkrecht", which translates to perpendicular; the "p" stands for the German word "parallel", which translates to parallel (that isn't a misprint, the German word for parallel is parallel). I like to use "s" for skip and "p" for pass-through.

1.3 (A Different Way to Find the Energy of a Photon)

(a) Using two formulas from Problem 1.2, derive a formula for the energy of a photon in terms of only fundamental constants and wavelength.
(b) Check to make sure the units are correct. Physicists always check units. Always.
   Hint: $1\,\mathrm{J} = 1\,\mathrm{kg\,m^2/s^2}$

1.4

(a) What is the energy of a photon that is in a laser that has a frequency of 647.8 THz (647.8×10^{12} Hz)? Express your answer in both joules (J) and electronvolts (eV). Your final answer should have 4 significant figures.
(b) What is the wavelength of the light in nanometers?

1.5 (Assessments) Physicists are trained to assess everything. It is what makes us special ☺.

Assess the formula you found in Problem 1.3(a) by re-finding the energy of a photon using the numbers from Problem 1.4(b). If you get the same answer as Problem 1.4(a), we have built confidence that the formula you derived is correct.

1.6 In atomic physics, energy is often measured in the units of $\mathrm{cm^{-1}}$. The method of calculating energy in these units is to first find the wavelength of the laser in centimeters, and then take the inverse. In equation form, this is

$$E\left(\mathrm{cm^{-1}}\right) = \frac{1}{\lambda(\mathrm{cm})}. \qquad (1.4)$$

Note that in the above formula, the units to use for each symbol are included in parentheses; the formula does not say "λ multiplied by cm", but rather "make sure wavelength has units of cm before plugging into the formula."

Using $\lambda = 852.347\,\mathrm{nm}$, find the energy in units of inverse centimeters. Your answer should have 6 significant figures.

1.7 The equation in Problem 1.6 can be confusing. Energy has units of joules and not inverse length! Why are atomic physicists comfortable with using energy in this weird unit? There are multiple correct answers here.

1.8 It isn't only atomic physicists who use energy in weird units. Astronomers, nuclear physicists, and a good number of chemists also use energy in inverse centimeters. Some chemists and condensed matter physicists prefer to use electronvolts (eV). Virtually no one uses the SI unit (joules)! Any thoughts why?

"Natural Light"

2

Abstract

In this chapter, we explore how light from sources such as the sun or a lamp can be dispersed into its spectral components. We discuss various dispersive elements, including gratings and prisms, and the phenomena of refraction and diffraction that allow these elements to spatially separate light. The chapter explores blackbody radiation and the historical significance of the ultraviolet catastrophe. Additionally, we introduce the concepts of absorption and emission lines, which provide insights into atomic and molecular energy levels.

Learning Goals

By the end of this chapter, you should be able to understand:

- the concept of spectral components and how white light is composed of various wavelengths of light.
- the function and importance of dispersive elements like prisms and gratings in spectroscopy.
- the difference between refraction and diffraction.
- blackbody radiation and the ultraviolet catastrophe.
- the principles of emission and absorption lines in spectroscopy and their significance in understanding atomic structure.
- the Stefan-Boltzmann law and Wien's displacement law in the context of blackbody radiation, and how to determine temperature from the blackbody spectrum.
- how light collected from the sun gave us our first evidence of atomic energy levels.

© The Author(s) 2025
W. Raven, *Atomic Physics for Everyone*,
https://doi.org/10.1007/978-3-031-69507-0_2

2.1   Breaking Light into a Spectrum

Definitions

- **Spectral component:** "White" light is made up of many different wavelengths of light. Even light from the sun, which looks yellow, is made up of many different wavelengths of light. A single wavelength of light that makes up a broader spectrum of light is called a spectral component.
- **Dispersive element:** Anything that spatially separates light into its spectral components.
- **Spectrometer:** Any tool that allows us to separate out and measure the amount of each spectral component.
- **Refraction:** the redirection of a wave as it passes from one medium to another medium (like air to water).
- **Diffraction:** when waves bend around the corners of an obstacle.

A spectrometer only needs two items: a dispersive element and a screen. A dispersive element is anything that takes light and spatially separates it into its spectral components. A common dispersive element is a prism, as shown in Fig. 2.1, which works due to refraction. If you send white light into a dispersive prism, you will see a rainbow exiting because each spectral component refracts at a different angle. For a prism, the smaller the wavelength of light, the larger the refraction angle. So, blue light ($\lambda_b = 400\text{--}490$ nm) refracts at a larger angle than red light ($\lambda_r = 620\text{--}750$ nm). If you calibrate the prism so that you know at what angle each

Fig. 2.1 A dispersive prism takes white light, which enters the prism from the left, and makes a rainbow. Each wavelength is refracted at a different angle. The white light source I used to make this picture is a tungsten lamp

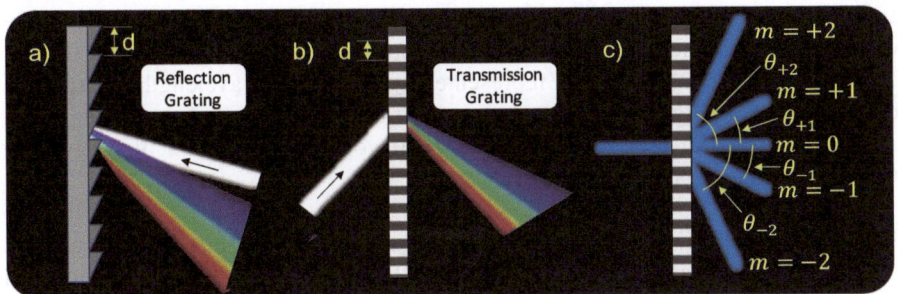

Fig. 2.2 (**a**) An example of white light, which is composed of many spectral components, reflecting off a reflection grating. For illustrative purposes, d, which is usually very small, is greatly enlarged. We also only show one order to keep the example a little cleaner. (**b**) An example of white light diffracting through a transmission grating. Again, d is greatly enlarged and we only show one order. (**c**) An example of blue light with $\lambda = 455$ nm hitting perpendicular to the transmission grating and being diffracted. In this example, we show all diffraction orders. The angles of diffraction are calculated using Eq. 2.1

wavelength refracts, you can send in an unknown wavelength, measure the angle of refraction, and use math to determine the wavelength of the light. You can also send in light from, for example, a hydrogen lamp and see what wavelengths or spectral components are in that light. Knowing how each wavelength refracts allows you to determine what wavelengths make up the hydrogen lamp spectrum.

In spectroscopy, the most common dispersive element is an optical grating, which works due to diffraction. Optical gratings commonly come in two types: transmission and reflection, see Fig. 2.2. Either way, the optical grating is a dispersive element that will spatially separate light into its spectral components because each spectral component diffracts at a different angle.

To create the diffraction, both types of gratings have small structures separated by a distance d. A reflection grating has a bunch of small tilted mirrors called rulings while a transmission grating has a bunch of small slits. The only physical criterion for a grating is that $d > \lambda$. The size of d in Fig. 2.2 has been greatly enlarged for visual purposes. While d must be larger than λ, in practice we also make sure that d is typically less than about 5λ to ensure suitable diffraction angles.

We will start with the equations that describe how spectral components are diffracted through a transmission grating; we will find that each spectral component can be diffracted at multiple, but well defined, angles. The math behind the next equation is a little complicated, but the end result is really what we care about. Let's start by keeping things simple and assume the incoming light is perpendicular to the grating, see Fig. 2.2c. The light diffracts through the grating according to the equation:

$$d \sin \theta_m = m\lambda$$
for incident light perpendicular to a transmission grating

(2.1)

In this equation, d is the distance between the slits, m is an integer known as the diffraction order (the fact that a spectral component diffracts at multiple but well defined angles is mathematically represented by m), λ is the wavelength of a spectral component of the light, and θ_m is the angle of diffraction for order m. m can be a negative integer, zero, or a positive integer. If you know d and m, you can then measure θ_m and use that to calculate λ.

Example A transmission grating has $d = 1\ \mu$m. There are three laser pointers: a blue laser pointer with $\lambda_b = 450$ nm, a green laser pointer with $\lambda_g = 532$ nm, and a red laser pointer with $\lambda_r = 650$ nm. All three lasers hit the grating perpendicular. What angles do the three lasers diffract for $m = +1, 0, -1$, and $+2$?

First, we use the diffraction equation to solve for θ_m: $\theta_m = \sin^{-1}(m\lambda/d)$, where $\sin^{-1}$ is the inverse sine function. For $m = 0$, we plug in numbers to find $\theta_{0,b} = 0°$, $\theta_{0,g} = 0°$, $\theta_{0,r} = 0°$. So, let's not use the $m = 0$ diffraction order since we can't learn anything here!

For $m = +1$, we plug in numbers to find $\theta_{+1,b} = 26.7°$, $\theta_{+1,g} = 32.1°$, $\theta_{+1,r} = 40.5°$. That is a pretty big difference! Your light only needs to travel a short distance to separate out the three colors.

Quick Math Aside
Let's put the screen 10 cm from the grating. On the screen is a ruler that will serve as our z-axis. If the grating wasn't in place, all spectral components would hit the screen at the same place. Let's call this spot $z = 0$ cm. With the grating in place and the white light hitting the grating perpendicular, the blue light, which diffracts at $26.7°$, hits the screen at $z = (10\text{ cm})\tan 26.7° = 5.04$ cm while the green light hits the screen $z = 6.28$ cm. You could also skip the angle calculation and use the formula:

$$z = \pm\frac{L}{\sqrt{(\frac{d}{m\lambda})^2 - 1}}$$

$$(2.2)$$

for incident light perpendicular to a transmission grating,

where L is the distance between the diffraction grating and the screen. z is positive for positive m and negative for negative m. You have the opportunity to derive this formula in problem 2.4. Notice that the diffraction order in the denominator is squared, so the sign of m only determines the sign of z.

For $m = -1$, we plug in numbers to find $\theta_{-1,b} = -26.7°$, $\theta_{-1,g} = -32.1°$, $\theta_{-1,r} = -40.5°$.

For $m = +2$, we find something interesting when we plug numbers into a calculator: $\theta_{+2,b} = 64.2°, \theta_{+2,g} = (90 - 20.4i)°, \theta_{+2,r} = (90 - 43.3i))°$. Blue light diffracts at 64.2°, but the other two have weird looking answers that came out of my calculator. These numbers are called complex numbers. Complex numbers are a huge topic in both math and physics, but what it means for us is that green and red light cannot diffract into the $m = +2$ order. Experimentally, we would see blue light diffracted at 64.2°, but we would not see second order diffraction for red light and green light. They just wouldn't be there. In general, whenever you calculate a number that is supposed to represent something physical and you get a complex number, that means this is something you cannot measure or it doesn't exist.

From this example, we would want to design our spectrometer to use either $m = +1$ or $m = -1$ because all spectral components diffract with a real, measurable angle (i.e. not a complex number). In the end, it doesn't matter which diffraction order we pick. As long as we know the angle at which a spectral component diffracts, we can use that information to determine the wavelength of an unknown spectral component.

A reflection grating does a similar job as a transmission grating, but the equation looks slightly different:

$$d \sin \theta_m = -m\lambda$$
for incident light perpendicular to a reflection grating
$$(2.3)$$

All that is different is the minus sign on the righthand side. Finally, let's suppose the incoming light isn't perpendicular to the grating but hits the grating at an angle θ_i with respect to the perpendicular (in optics, the perpendicular is called the "normal"), see Fig. 2.2b. The formula describing the diffraction of a spectral component is:

$$d(\sin \theta_m - \sin \theta_i) = \pm m\lambda$$
(use $+ m$ for transmission and $- m$ for reflection)
$$(2.4)$$

While this formula is more complicated, it is important to note that the concept is far more important than the formula. The concept is simply this:

Take Home Message
If you know how your dispersive optic bends (diffracts or refracts) different wavelengths of light, you can use that information to determine the wavelength of an unknown spectral component from any light source.

Fun History

- The first known exploration of a diffraction grating was by the Scottish mathematician and astronomer James Gregory in the mid 1600s. He observed the diffraction of sunlight caused by light passing through a bird feather. The individual feathers acted like the small slits of a transmission grating. The German physicist Joseph von Fraunhofer made the first diffraction grating in 1814. He discovered something amazing, which is the topic of the next section.
- Diffracting light from a light bulb whose gas is a particular element like hydrogen, nitrogen, or oxygen is the topic of Sect. 2.3. Classical physics had no way of explaining the observed spectral components, and this mystery was one of the puzzling experiments that led to quantum mechanics.

2.2   Blackbody Radiation

I pointed a commercial spectrometer at the sky, see Fig. 2.3. Collecting this data was spectroscopy! A few important notes before we continue:

- This spectrometer has a dispersive element that was pre-calibrated, so the angle the light hits the detector is automatically converted to a wavelength.
- Spectrometers are not uniformly sensitive. This means that the detector you are using might be more sensitive to red light than blue light. Graphs like Fig. 2.3 are useful for identifying the wavelengths present in the light, but unless the detector has been calibrated to correct for wavelength sensitivity, the shape may

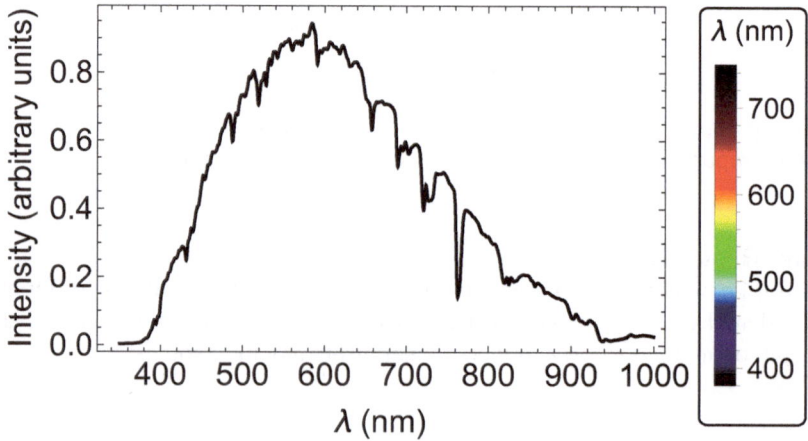

Fig. 2.3 The spectral components of light from the sun after it passes through the atmosphere

be incorrect. In fact, the real spectrum of the sky goes to far longer wavelengths, but this spectrometer is not sensitive to light at those frequencies.

- When this type of spectrum was first taken in the 1800s, those dips were a mystery, and physicists love a good mystery. Whenever a physicist is presented with data, the first thing they ask themselves is, "Why does the data look the way it does?" If we understand the concepts, we should be able to produce a mathematical model to describe (and predict) the data.
- For this data, there are, at least, two things that need to be thought through. The first is the overall shape. The second is, what are those dips?

Historically, it was a fun journey to figure out the overall shape. Physics known in the 1800s predicted a far different shape. In fact, when challenged with describing the overall shape, physics failed hard. When physicists tried to predict the spectrum in Fig. 2.3, the model was close for large wavelengths but very, very wrong for small wavelengths. Starting at the larger wavelengths and moving to shorter wavelengths, the math says the spectrum should get larger and larger and larger instead of turning around and getting smaller, which would require infinite energy. The dramatic result of the math is this: the universe doesn't exist. Whenever the model says the universe doesn't exist, there is something we don't understand.[1] This result is now known as the "ultraviolet catastrophe." It is a dramatic name, but the predictions from the mathematical model were also dramatic. We clearly needed a better model!

A Short Aside

Imagine you had a room with perfectly reflecting, parallel walls. Only light with particular wavelengths that create standing waves (perfect constructive interference like in Fig. 1.7) can exist in this room. Even though only certain wavelengths are allowed in the room, there are still an infinite number of them (you can always add one more loop to the standing wave). According to classical thermodynamics, a well tested and very successful theory, each standing wave has the same amount of energy. If you learned about heat capacity in high school chemistry, this classical thermodynamics model predicted an infinite heat capacity. Yikes!

According to the model, since each mode has the same energy, Fig. 2.3 should keep going up and up at small wavelengths. As a practical example, this mathematical model says that if you stood in a closed room and lit a match, the entire room and everything in it would burst into flames due to the infinite energy density at short wavelengths. Ultraviolet catastrophe indeed!

---

[1] I feel like this might be the understatement of the millennium.

In 1901, after a lot of thought and model development, German physicist Max Planck eventually figured out that if light was composed of photons, then there would be way fewer higher-energy photons than lower-energy ones, which solved the problem. In 1905, German born physicist Albert Einstein built upon this idea with the photoelectric effect. The mathematical result of this new model predicted the shape in Fig. 2.3, which is modeled by the complicated formula:

$$M(\lambda, T) = \frac{2\pi h c^2}{\lambda^5} \frac{1}{e^{\frac{hc}{\lambda k_B T}} - 1}, \tag{2.5}$$

where $h = 6.626 \times 10^{-34}$ Js is Planck's constant, c is the speed of light, $k_B = 1.38 \times 10^{-23}$ J/K is a constant known as Boltzmann's constant,[2] and T is the temperature of the object. For temperature, we use the unit kelvin, which is named after British mathematician, physicist, and engineer Lord William Thomson. The unit for Boltzmann's constant is joule/kelvin. Using this model to fit the data from the spectrometer, we can measure the temperature of the surface of the sun to be about 5800 kelvin. We call $M(\lambda, T)$ "spectral radiant exitance", and it tells us how much radiant energy per second is leaving the object per unit area per unit wavelength.[3] The units for $M(\lambda, T)$ are W/m^3. I should note that this equation is for a "perfect" blackbody, which doesn't exist in real life. Not having a perfect blackbody is equivalent to saying there is no such thing as a room with perfectly parallel and reflective walls. However, it still does a pretty good job with modeling the spectrum emitted by objects.

It isn't just the sun that emits a blackbody spectrum. A hot pan on the stove does, you do, and I do as well. The amount of light that is emitted as well as the peak wavelength of the blackbody spectrum only depends upon the temperature of the object. Remarkably, everything else in this model is either a constant or wavelength, which is our horizontal axis. Figure 2.4 shows a few different plots of the spectral radiant exitance for different temperatures. For objects near 5000 K, like the sun, they emit light that is visible to our human eye. For cooler objects, like us, the maximum spectral radiant exitance occurs more near 10,000 nm, which is 10 μm. There are special cameras that can see this wavelength of light. You may have seen thermal imaging or watched the Predator movies. I did not include the vertical scale in this plot because the numbers are big and hard to interpret without really digging into the spectral radiant exitance formula. But hotter objects do emit more light. The maximum spectral radiant exitance for a 6000 K object is about 3 million times larger than for a 310 K object.

---

[2] Boltzmann's constant was named by Max Planck in honor of the Austrian physicist and philosopher Ludwig Boltzmann.

[3] The units are, admittedly, confusing! As a rough interpretation, you can think of these units as telling you how much light is being emitted from a blackbody at a given wavelength. If you look at Fig. 2.3, there is more light being emitted at 500 nm than 700 nm, so the spectral radiant exitance is larger at 500 nm. This is what we mean by "per unit wavelength."

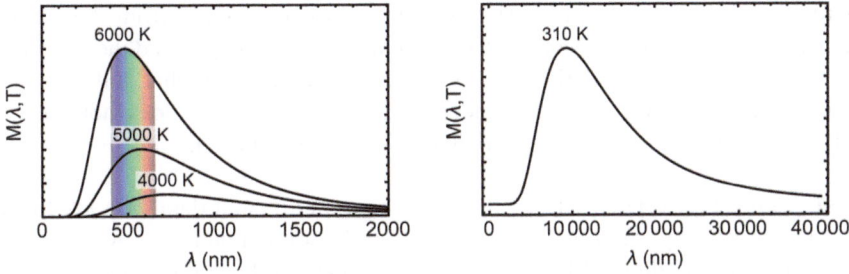

Fig. 2.4 Left: Blackbody spectrum for very hot objects. The surface of the sun is 5800 K. Right: Your blackbody spectrum, assuming you are human (310 K is about 98 F). Notice the horizontal axis is a very different scale

The Stefan–Boltzmann Law and Wien's Displacement Law Imagine you have a blackbody with surface area A. Besides the spectrum of the blackbody, you can also measure the total power emitted by the object. The total power emitted is proportional to the area under a spectral radiant exitance graph. After some math (calculus), the total power emitted by a blackbody:

$$P = \sigma \epsilon A T^4,$$
$$\sigma = \frac{2\pi^5}{15} \frac{k_B^4}{c^2 h^3} = 5.67 \times 10^{-8} \, \frac{W}{m^2 K^4} \tag{2.6}$$

where σ is a constant known as the Stefan–Boltzmann constant,[4] ϵ is the emissivity, and A is the surface area of the object. Emissivity is a number between 0 and 1 indicating how closely the object behaves like a blackbody. If it is a perfect blackbody, $\epsilon = 1$. For a sphere, like the sun, the surface area is $4\pi R^2$, where R is the radius of the sphere. The emissivity of the sun is about 0.99, so it is nearly a blackbody.

Several years before Max Planck solved the ultraviolet catastrophe, German physicist Wilhelm Wien experimentally noticed that the wavelength of the maximum spectral radiant exitance graph, see Fig. 2.4, was inversely proportional to temperature. He discovered that a blackbody with a temperature of 6000 K had a maximum around $\lambda_{peak} = 484$ nm, 5000 K had a maximum around 580 nm, 4000 K around 725 nm, etc. After studying the trend between temperature and peak wavelength, he determined:

$$\lambda_{peak} = \frac{b}{T}, \tag{2.7}$$

---

[4] Named after the Carinthian Slovene physicist, mathematician, and poet Josef Stefan, who empirically found the relationship, and Austrian physicist and philosopher Ludwig Boltzmann, who derived the equation.

where $b = 2.898 \times 10^6$ nm K. This formula was later derived from the spectral radiant exitance formula. If you have the calculus background to find the maximum of a function and want to derive this formula, you should know that, unfortunately, the resulting equation is a transcendental equation and thus does not have an analytical solution. However, you can find a numerical solution.

What About Those Dips? Let's return to the experimental data we took for the spectrum of the sky, see Fig. 2.3. Blackbody radiation explains the overall shape, but what about the dips? What do you think is making those dips? The answer is below the fun little puzzle in Fig. 2.5.

Fig. 2.5 A fun little puzzle to separate the question from the answer

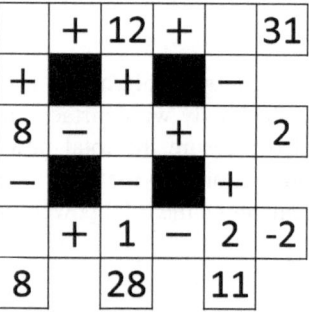

There are two sources of the dips: atoms and molecules in our atmosphere and the sun itself. For example, the dip near 750 nm is due to oxygen molecules in the atmosphere. There are photons coming from the sun that have the perfect energy to excite oxygen molecules. Because oxygen molecules absorb only photons with the perfect energy, some light is lost at very specific wavelengths before it reaches our spectrometer.

The second source for the dips is the sun itself. While the sun emits blackbody radiation, that light passes through gas at the surface of the sun. That gas absorbs light just like atoms and molecules in our atmosphere. If we wanted just the composition of the sun, we should collect this data from space!

This leads to one method of performing spectroscopy with a spectrometer. You take a light source, which can be a blackbody, a flashlight, or really any light source you want, and send it into your spectrometer. Record this spectrum. Next, place an atomic sample between the light source and the spectrometer and record this spectrum. Comparing the two spectra tells you which wavelengths got absorbed. Each wavelength that is lost from the light due to absorption is the same wavelength needed to excite the atom from one energy level to another. So measuring these absorption dips gives you information about the energy levels of the atoms.

2.3   Discharge Lamps

Another method of spectroscopy using a spectrometer is to point the spectrometer at a lamp filled with an element. Figure 2.6 shows the spectrum of a helium discharge lamp collected using a commercial spectrometer. The spectrum is not continuous like that of a blackbody. Instead, we observe individual spikes at very specific wavelengths. And, we already know the source of these spikes! Atoms have energy levels, and the atoms will only absorb and emit at specific wavelengths. Filling a lamp with a specific element, or the combination of a few elements, and collecting the emitted light with a spectrometer was the most common form of spectroscopy before the invention of the laser. After the spectrum was collected, physicists had to be very clever to back out the energy levels from all of those lines.

Whether you use a white light source and look for lost photons or you collect light from a discharge lamp, you gain the same information, which is the wavelengths of light needed to have an atom go from one energy level to another.

In the lab, spectrometers are still really useful tools, but they are not great for certain applications. Below are a few pros and cons to using spectrometers for spectroscopy.

Pros:

–   You get lots of data all at once.
–   You get lots of data very quickly.

Cons:

–   The resolution is limited. You need to get different wavelengths far enough apart to distinguish between them. For example, let's see how far apart 450.334 nm is

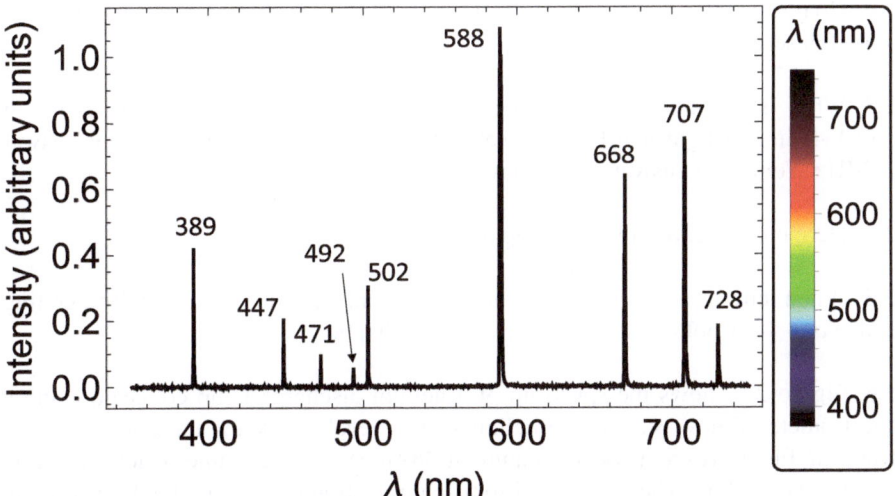

Fig. 2.6 Spectrum collected from a helium discharge lamp collected using a spectrometer with an accuracy of 0.5 nm

from 450.335 nm using Eq. 2.2 with $L = 10$ cm, $m = +1$, and $d = 1000$ nm:

$$z(\lambda) = \pm \frac{L}{\sqrt{(\frac{d}{m\lambda})^2 - 1}}$$

$$z(450.335\,\text{nm}) - z(450.334\,\text{nm}) = 140\,\text{nm}.$$

That is small! A possible solution is to make a bigger spectrometer, but then you will have to worry about thermal drifts, how well you can uniformly space the grating slits, and how well you can measure d. In short, it becomes really hard if you want really high resolution.

– The data can be hard to interpret. Getting all the data at once means you are getting all transitions from your atom or molecule all at once. A spectroscopist has to figure out what line comes from what transition (pair of energy levels), which is not an easy task!

For high precision spectroscopy, most spectroscopy groups use a laser to study a single transition instead of collecting light from many transitions at once. As we will find in Chap. 4, having atoms moving around (i.e., a hot gas) is a bad thing if we want to measure something like the energy difference between two energy states to really high precision. Clever physicists developed a neat technique to still use hot atoms to measure properties to high precision, which is the topic of Chap. 5, but we need multiple lasers to do so.

Problems

2.1 Go through the chapter and write down all of the new fundamental constants. Don't forget units.

2.2 Go through the chapter and write down each equation. For each equation, write a brief description about what the equation means.

2.3 What is the frequency difference between light that has a wavelength of 450.334 nm and light that has a wavelength of 450.335 nm? Express your answer in MHz. Your final answer should have 6 significant figures.

2.4 Derive Eq. 2.2. Start by drawing a triangle.

2.5 When introducing gratings we stated, "The only physical criterion for a grating is that $d > \lambda$." Looking at Eq. 2.2, what would happen if $d < \lambda$?

2.6 Figure 2.6 shows the spectrum of a helium discharge lamp collected using a spectrometer. Below is a table of helium energy levels. Pick any three spectral lines from the Figure (except for the feature at 389 nm) and determine which transition produced each line. The lower level for each transition is listed in the third column.

For example, the feature at 389 nm is due to an electron transitioning from level 8 to level 2:

$$185, 564.6 \, \text{cm}^{-1} - 159, 856.0 \, \text{cm}^{-1} = 25708.6 \, \text{cm}^{-1}$$
$$\rightarrow \lambda = \frac{1}{25708.6 \, \text{cm}^{-1}} = 0.0000388975 \, \text{cm} = 388.975 \, \text{nm}$$

Level #	Energy (cm^{-1})	Lines with this lower level
1	0	
2	159,856.0	389 nm
3	166,277.4	502 nm
4	169,086.8	447, 471, and 707 nm
5	171,134.9	492 and 728 nm
6	183,236.8	
7	184,864.8	
8	185,564.6	
9	186,209.4	
10	190,298.1	
11	191,446.5	

2.7 You want to design your own spectrometer to measure light collected from a lamp. You need to collect data for light with wavelengths between 1000 and 3000 nm. You have a number of different gratings that you can pick from, each with a different d. What are a few examples of a bad choice for d? What is an example of a good choice for d?

2.8 (Solar Power)
The sun has a surface temperature of 5772 K with a radius of $r_s = 696, 340 \, \text{km}$. The surface area of the sun is $A = 4\pi r_s^2$. The emissivity of the sun is about 0.99.

(a) Find the total power output of the sun.
(b) The light from the sun spreads out radially in all directions. A small fraction of that light hits the earth. The earth is $d = 1.496 \times 10^8 \, \text{km}$ from the sun. Assuming the earth is a solid disk with radius $r_e = 6371 \, \text{km}$, what fraction of the total power leaving the sun hits the earth?
Hint: We first want to find the ratio of the area of the earth disk (πr_e^2) to the surface area of a sphere with the radius equal to the earth-sun distance: $\frac{\pi r_e^2}{4\pi d^2}$
That ratio tells us fraction of light emitted from the sun that hits the earth.
(c) For solar energy, we care about how much power per area is hitting the solar panel. Power per area is called intensity. Find how much power from the sun is hitting a 1 square meter area of land. This is the intensity of sunlight on the earth.
(d) You have a 10 cm by 10 cm solar panel. Determine the power of sunlight hitting that solar panel.

(e) Your solar panel is 20% efficient at converting sunlight into usable electric power. What is the power output of your solar panel? Is it enough to power a 10 watt LED light bulb?

(f) A refrigerator needs 200 watts of power to run. What area solar panel do you need to run the refrigerator? If the solar panel was a square, what are the dimensions of that square?

2.9 (Numerical Problem)

(a) Figure 2.7 is an amazing figure made by Robert A. Rohde. It shows the spectrum from the sun collected using an amplitude corrected spectrometer. The vertical axis is the spectral irradiance.[5] Included in the graph is a good approximation of what the sun would emit if it were a perfect blackbody. Using your favorite graphing program, estimate the temperature of the surface of the sun by plotting the spectral radiant exitance. Don't worry about the vertical scale. What you are most concerned about is getting the spectral radiant exitance to be a maximum around 500 nm.

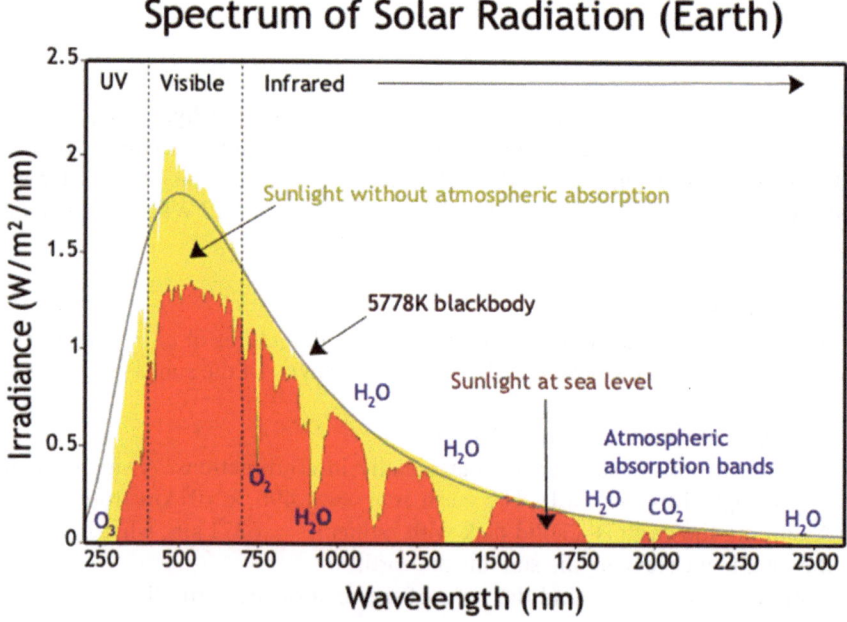

Fig. 2.7 The spectrum from the sun both before the light enters the atmosphere (yellow) and at the surface of the earth (red). A perfect blackbody spectrum is shown by the black curve. Image Credit: Robert A. Rohde, CC BY-SA 3.0 via Wikimedia Commons

---

[5] Spectral radiant exitance is the radiant flux *emitted* by a surface per unit area per unit wavelength. Spectral irradiance is the radiant flux *received* by a surface per unit area per unit wavelength. You can think of spectral radiant exitance as what leaves the sun and spectral irradiance as what hits the earth.

(b) Use Wien's displacement law to assess your answer. (In other words, does Wien's displacement law confirm the peak of your graph in part (a)?)

2.10 (Advanced Problem)

(a) Using your favorite numerical program, numerically find and plot the wavelength of maximum spectral wavelength (λ_{peak}) for temperatures between 3000 and 6000 K in steps of 100 K.[6]
(b) Fit your data to b/T to find b. Be sure to use the speed of light accurate to, at least, 5 digits.

2.11 (Advanced Math Problem) Requires calculus.

(a) Find the transcendental equation that you would need to numerically solve to find Wien's displacement law.
(b) Convert the spectral radiant exitance from a function of wavelength to a function of frequency. To do this, you need to use the formula. $M_\lambda(\lambda, T)d\lambda = -M_f(f, T)df$, which guarantees that the same amount of total energy is in a spectral interval $d\lambda$ as in the corresponding interval df. The minus sign is because decreasing wavelength increases frequency.
(c) Find the spectral radiant exitance as a function of wavenumber (inverse wavelength).

---

[6] In python, you can use the code scipy.optimize.fmin from the scipy library. The code only finds the minimum, so you would multiply the spectral radiant exitance formula by -1 first. In Mathematica, the function is FindMaximum.

Atoms at Rest

3

Abstract

In this chapter, we consider the factors that lead to complexity in atomic lines. We will learn that all spectral lines have a fundamental (natural) width and exist as a spread of frequencies rather than a single frequency. The width of atomic lines can also be affected by external factors such as pressure and laser power. This will be related to what we observe when probing an atom with a laser. In addition, we will learn about how this fundamental width is related to the lifetime of a state, how many photons per second an atom can absorb from a laser, and the two major types of experimental plots known as absorption plots and transmission plots.

Learning Goals

By the end of this chapter, you should be able to understand:

- the relationship between the natural linewidth of a transition and the lifetime of a state.
- absorption plots and transmission plots.
- the scattering rate and its influence on the absorption and transmission plot.
- the saturation intensity of a transition.
- the saturation parameter.
- power broadening.

© The Author(s) 2025
W. Raven, *Atomic Physics for Everyone*,
https://doi.org/10.1007/978-3-031-69507-0_3

3.1   A Thought Experiment

An important statement and two definitions from Chap. 1:

▶   **Important Statement** If a photon has the same energy as the energy
     difference between the ground state and an excited state, the atom will
     absorb that photon and move an electron to the excited state. If the
     photon does not have the same energy as that energy difference, the
     atom will ignore the photon completely.

Definitions

- **Resonance:** When an atom gets excited by a photon from one state to
  another, we say the atom "goes through resonance." This is similar to
  playing the trumpet. When you blow into a trumpet, you excite a standing
  wave in the pipe to create a note. The same thing happens with an atom.
  When you excite an atom, the electron goes from one standing wave mode
  to another. Atomic physicists use the word "excitation" and "resonance"
  interchangeably.
- **Resonance frequency:** When the frequency of the laser is just right to excite
  the atom from the ground state to an excited state, we call that the resonance
  frequency. We use the variable f_r for resonance frequency. Since resonance
  frequency is a frequency, it uses units of Hz, MHz, GHz, or THz.

Imagine we have a bunch of atoms in a vapor cell, which is a sealed glass
tube containing only the atoms we are interested in. We are going to make a few
assumptions:

- The atoms are in their gaseous form.
- All of the atoms are frozen in place. In other words, the speed of every atom is 0
  m/s (at rest).
- None of the atoms are interacting with anything, including each other. This
  means that all the atoms are neutral (each atom has an equal number of electrons
  and protons)
- Every atom has only 2 energy states. The state with lower energy is called
  the ground state. The state with higher energy is called the excited state. This
  simplified atom has a very descriptive label: the two-level atom.

Now, we do spectroscopy. We start by sending a laser through the vapor cell and
detecting how much light makes it through, see Fig. 3.1. At the beginning of the
thought experiment, the photons in the laser do not have enough energy to excite
the atoms from the ground state to the excited state. In other words, the frequency
of the laser is lower than the resonance frequency. As such, we don't expect the

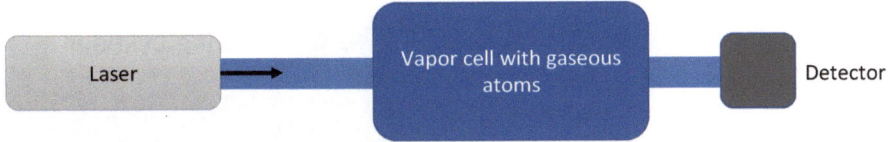

Fig. 3.1 The experimental setup for the thought experiment

atoms to absorb photons from the laser beam. Next, we will smoothly increase the frequency of the laser until the photons in the laser beam have more energy than the energy difference between the ground and excited states. In going from too small to too big, the laser frequency will, at some point, be just right so that the photons have the right amount of energy to excite the atom, which we call resonance. In the lab, smoothly changing the laser frequency over time is referred to as "scanning the laser."

A transmission plot is a plot of the fraction of photons that make it through the vapor cell as a function of laser frequency. The question is, what does our transmission plot look like for this thought experiment? Based on what we've learned so far, a completely reasonable guess would be that the atoms completely ignore the photons unless the photons have the perfect energy to excite the atom from the ground state to the excited state. So, you might guess that our transmission plot looks like the sharp dip in Fig. 3.2. The plot has a single, sharp dip that occurs at the resonance frequency. If the laser has any other frequency the photons do not have the correct energy to excite the atom. However, this is not quite right. A more realistic transmission plot can be seen in Fig. 3.3a. The transmission plot does indeed have a dip at the resonance frequency, but the dip has a width. This width is

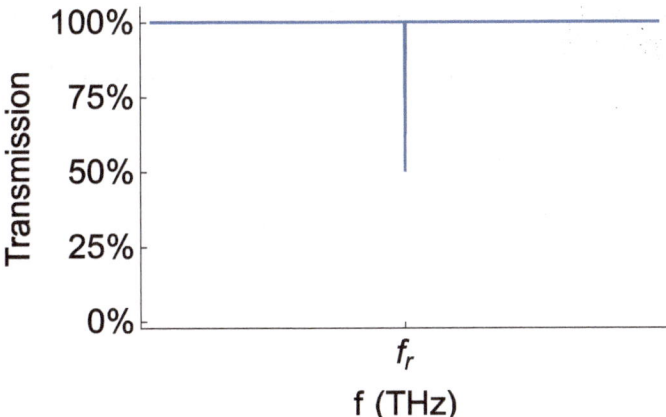

Fig. 3.2 A completely reasonable, but incorrect, guess for the transmission plot. For a transmission plot, 100% means that no photons are absorbed by the atoms, 50% means that half of the photons are absorbed, and 0% means that all of the photons are absorbed by the atoms

a)

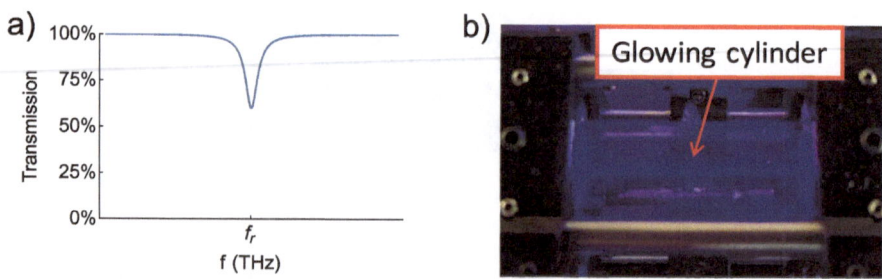

Fig. 3.3 (a) A more accurate transmission plot for the thought experiment. (b) A picture of a laser beam whose frequency matches a resonance frequency near 459 nm (blue light) passing through a vapor cell of cesium atoms. As we scan the laser frequency from below the resonance frequency to above the resonance frequency, we visually see no glowing cylinder, followed by a glowing cylinder (resonance), followed by no glowing cylinder. On resonance, the transmission decreases as shown in (a) because the atoms take photons from the laser and re-emit them in all directions, and almost none of these re-emitted photons continue along their original path to the detector. Note that the transmission plot in (a) is a theory plot assuming that the atoms are at rest. In the vapor cell the atoms are not frozen in place. Chapter 4 will explore a transmission plot for atoms moving around

called the **natural linewidth** of the transition and is represented by the lowercase Greek letter gamma, γ. The natural linewidth is a frequency, so it has units of hertz. Spectroscopists often just say "linewidth" instead of "natural linewidth." In an experiment, we would see our atoms start to glow as the laser frequency passes through resonance, see Fig. 3.3b.

Often times spectroscopists would rather look at an absorption plot instead of a transmission plot, see Fig. 3.4. An absorption plot is the fraction of photons lost as a function of laser frequency. It looks very similar to a transmission plot, but it has a bump instead of a dip. Both an absorption plot and a transmission plot tell us the same information: atoms absorb (and re-emit) photons from the laser around the resonance frequency. If you add absorption to transmission, you should get 100% for all laser frequencies.

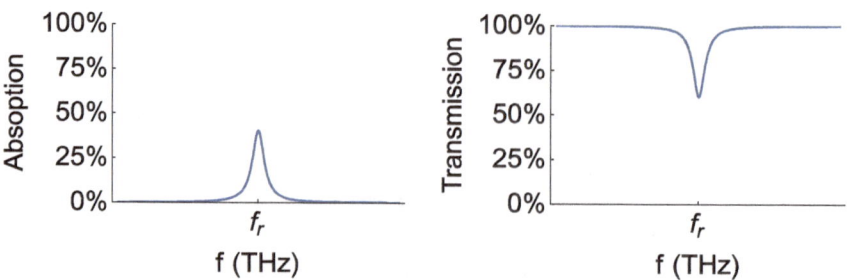

Fig. 3.4 An illustrative example of an absorption plot (left) and a transmission plot (right). If you add these two plots together, you would get 100% for all laser frequencies

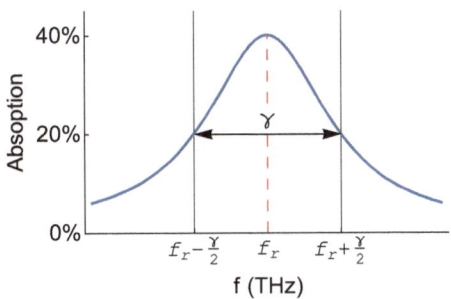

Fig. 3.5 A zoom in of a spectral feature. The natural linewidth is the full width half maximum of this feature

The natural linewidth is the full width at half the maximum (FWHM) of the absorption bump, see Fig. 3.5. It is a property of the transition that we cannot change. As an analogy, think about the charge or mass of an electron. The charge of the electron is simply the charge of the electron, which is 1.602×10^{-19} coulombs. The mass of the electron is simply the mass of the electron, which is 9.11×10^{-31} kg. These are intrinsic properties of the electron that we cannot change. The natural linewidth of a transition is inherent for that transition, and we cannot change it.

The shape of the bump (or dip in the transmission plot) is often referred to as a **spectral feature** or spectral profile. For completeness, the width of the spectral feature we measure in the lab is always larger than the natural linewidth because of various "broadening" mechanisms. One of these broadening mechanisms is laser power, which we will discuss in Sect. 3.5. If the laser power was the only broadening mechanism, we would find that as the laser power gets smaller and smaller, the width gets narrower and narrower until it reaches the natural linewidth. The natural linewidth is the *minimum* possible FWHM of a spectral feature.

The mathematical shape of the spectral feature is a Lorentzian function. In the absence of any broadening mechanism, the mathematical form is:

$$L(f) = \frac{A}{1 + \frac{4(f - f_r)^2}{\gamma^2}}, \tag{3.1}$$

where γ is the natural linewidth[1] and A is the maximum absorption, which describes the amount of laser light lost when traveling through an atomic sample when the laser light is perfectly on resonance. A is a number between 0 (no light absorbed) and 1 (all light absorbed, which is also 100%). You may be familiar with a similarly shaped Gaussian function. A Lorentzian function looks very similar, but it is slightly different in shape.

---

[1] This is for the low power limit. In Sect. 3.5, we will refine this formula slightly to include power broadening. I just want to start basic to get the concepts first.

Math Assessment

Equation 3.1 is the mathematical shape of a spectral feature. According to Fig. 3.5 (and the definition of natural linewidth), the absorption at $f = f_r \pm \gamma/2$ should be half as large as the absorption when $f = f_r$. So, let's check to make sure the formula matches that statement. To do this, we will first plug in $f = f_r$ into Eq. 3.1 to make sure that $L(f_r) = A$:

$$L(f_r) = \frac{A}{1 + \frac{4(f_r - f_r)^2}{\gamma^2}}$$

$$= \frac{A}{1 + \frac{0}{\gamma^2}} = A$$

Next, we will plug in $f = f_r \pm \gamma/2$ to see if $L(f_r \pm \gamma/2) = A/2$. If you are more comfortable doing this twice, once for $f = f_r + \gamma/2$ and once for $f = f_r - \gamma/2$, please do so!

$$L(f_r \pm \gamma/2) = \frac{A}{1 + \frac{4(f_r \pm \gamma/2 - f_r)^2}{\gamma^2}}$$

$$= \frac{A}{1 + \frac{4(\pm\gamma/2)^2}{\gamma^2}} = \frac{A}{1 + \frac{4(\gamma^2/4)}{\gamma^2}} = \frac{A}{1 + \frac{\gamma^2}{\gamma^2}} = \frac{A}{1+1}$$

$$= \frac{A}{2}$$

Yay!

Important Spectroscopists being spectroscopists call the spectral feature many different things. Other common descriptions of the spectral feature include the Lorentzian profile, the absorption profile, the absorption lineshape, the spectral profile, the spectral lineshape, or the Doppler-free spectrum. The description "Doppler-free spectrum" will make more sense after Chap. 5. Some atomic physicists use the descriptor "Lorentzian function", but that should be avoided because that phrase is a mathematical definition of that function itself. As an analogy, think about how many times you have used a sine wave or a cosine wave in math. These are mathematical functions that are used to describe lots of different physical phenomena like an oscillating spring or a swinging pendulum. The same thing is true with the Lorentzian function. There are other physical phenomena described by a Lorentzian function, so we instead use more specific language. For the rest of the book, we will use the term spectral feature or spectral profile.

Definitions

- **Transmission plot:** A plot of the percentage or fraction of photons that passes through a vapor cell as a function of laser frequency.
- **Absorption plot:** A plot of the percentage or fraction of photons lost from a laser after it passes through a vapor cell as a function of laser frequency. A dip in a transmission plot is seen as a bump in the absorption plot.
- **Natural linewidth:** The minimum possible full width at half maximum (FWHM) of a spectral feature. The natural linewidth is a property of a transition, with each transition in an atom having a unique natural linewidth.

One Last Thing We can now update that important statement from Chap. 1, which is also at the top of this section.

▶ **Important Statement** A photon has a probability of being absorbed by an atom depending upon the photon's energy. It is most likely to be absorbed if the photon's energy exactly matches the energy difference between the ground and excited state, but there is a non-zero probability of absorption off resonance.

Specifically, in the absence of any broadening mechanism (like laser power), if the photon's energy is off by $\frac{h\gamma}{2}$ from that resonance energy, the photon is half as likely to be absorbed compared to a photon that has the energy equal to the energy difference between the ground and excited states.

3.2 The Natural Linewidth in Angular Units

Many formulas that spectroscopists and atomic physicists use contain variables with angular units. This is more of a mathematical convenience, but since they are used so often, I wanted to introduce them as a topic. Suppose we had a simple sine wave that described an oscillation in time with frequency f and amplitude A. We would write that as:

$$A \sin (2\pi f t) \tag{3.2}$$

The presence of 2π in the argument of the sine function stems from the inherent periodicity of a sine wave, which repeats itself every 2π radians. Consequently, each time the argument of the sine function increases by 2π, the waveform completes one full cycle. Thus, the frequency denotes the rate at which the sine wave repeats within a time span of one second. This periodicity of 2π is where the "angular" part comes in. Instead of always writing $2\pi f$, we simplify things by using angular frequency,

$\omega = 2\pi f$. That curly-looking symbol is the lowercase Greek letter omega. That same sine wave using angular frequency units is:

$$A \sin (\omega t). \tag{3.3}$$

In the lab, we measure frequency, but, mathematically, we often use angular frequency. The natural linewidth is a frequency, but it is often written in formulas using angular frequency units. In this book, we will use capital gamma, Γ, for the natural width in angular frequency units and lowercase gamma, γ, for normal frequency units. The relationship between the two is:

$$\Gamma = 2\pi \gamma. \tag{3.4}$$

I want to emphasize that frequency, and not angular frequency, is the unit that we work with in the lab. To help us keep these two parameters separate, we use the unit radians/second $\left(\frac{rad}{s} \right)$ for angular frequency and inverse seconds $\left(\frac{1}{s} \right)$ or hertz (Hz) for frequency. If you are given the information that $\Gamma = 10.5 \times 10^6 \frac{rad}{s}$, then we calculate $\gamma = \frac{10.5 \times 10^6 \, rad/s}{2\pi} = 1.67 \times 10^6 \frac{1}{s} = 1.67\,MHz$ for use in the lab. If we want to change the frequency of a laser by one linewidth, we change the frequency by 1.67 MHz, not 10.5 MHz. Again, to avoid confusion, we don't use the unit MHz for the angular frequency. Instead, we make sure to use rad/s: $10.5 \times 10^6 \frac{rad}{s}$.

Reduced Planck's Constant In the explanation of the photoelectric effect, Einstein discovered that a photon has an energy $E_{ph} = hf$. If we wanted to write this equation using angular units, we would get $E_{ph} = h\frac{\omega}{2\pi}$. Physicists absorb that extra 2π into h and call that new constant the "reduced Planck's constant": $\hbar = \frac{h}{2\pi} = 1.054 \times 10^{-34}$ Js. Notice that the variable for the reduced Planck's constant resembles a lowercase h with a horizontal line drawn across its stem. We call this constant "h-bar". We now have two completely equivalent ways of writing the energy of a photon:

$$E_{ph} = hf \text{ (frequency units)}$$
$$E_{ph} = \hbar\omega \text{ (angular frequency units)} \tag{3.5}$$

It doesn't matter what equation you use, you will always calculate the same energy.

Any function expressed in frequency units can be written in angular frequency units. For example, the Lorentzian function from Eq. 3.1 using angular frequency units is:

$$L(\omega) = \frac{A}{1 + \frac{4(\omega - \omega_r)^2}{\Gamma^2}}, \tag{3.6}$$

where $\omega_r = 2\pi f_r$ is the angular resonance frequency.

3.3   The Natural Linewidth and the Lifetime of the Excited State

The natural linewidth is, remarkably, related to the lifetime of the excited state. To understand the lifetime of an atomic state, first imagine that all of the atoms from the above thought experiment are in the excited state. Next, we turn off the laser beam. As time progresses, the atoms will each emit a photon to go back to the ground state at a random time. Spectroscopists often use the word "decay" when talking about an atom emitting a photon to go back to the ground state. This is a random process, so some atoms decay back to the ground state quickly while others take their time. Statistically, each atom has a probability of decaying out of the excited state that will look like Fig. 3.6. At $t = 0$, all the atoms are in the excited state. As time marches on, atoms start to decay and the excited state fraction gets smaller. For the graph in Fig. 3.6, about 63% of the atoms have decayed back to the ground state in 6.25 ns, and about 80% of the atoms have decayed back to the ground state in 10 ns. The mathematical form of the decay curve is:

$$e^{-t/\tau}, \tag{3.7}$$

where τ (this is the Greek letter lowercase tau) is called the lifetime of the excited state. This function is called an exponential decay. Like the natural linewidth of a transition, τ has a unique value for every transition in an atom. The lifetime of the excited state used for the decay in the graph is $\tau = 6.25$ ns, which means that after 6.25 ns about 63% of the atoms have decayed. There is nothing special about 63%; it is just what physicists decided to define as the lifetime. Mathematically it is a nice definition because when $t = \tau$, the fraction of atoms that have not decayed is $e^{-1} = 0.368$, or 36.8%. After two lifetimes (2×6.25 ns $= 12.5$ ns), about 86.5% of the atoms have decayed leaving $e^{-2} = 0.135$, or 13.5%, of the atoms in the excited state.

Amazingly, the natural linewidth and the lifetime are related! The formula relating the two quantities is:

$$\tau = \frac{1}{\Gamma} = \frac{1}{2\pi\gamma}. \tag{3.8}$$

Fig. 3.6 The excited state fraction as a function of time for an excited state that has a lifetime of 6.25 ns

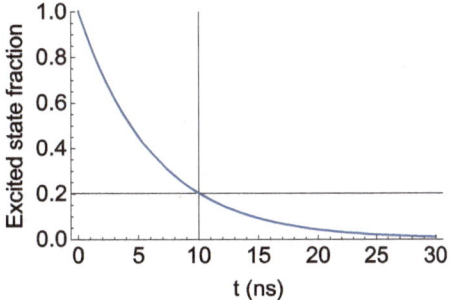

As you can imagine, atomic physicists and spectroscopists often use the angular frequency form of natural linewidth when thinking about lifetime.[2] These two quantities are inversely related to each other. This means:

- If the lifetime is small, the natural linewidth is large. This means that excited states with a small lifetime have absorption and transmission plot spectral features that are relatively wide.
- If the lifetime is large, the natural linewidth is small. This means that excited states with a large lifetime have absorption and transmission plot spectral features that are relatively narrow.

3.4   The Scattering Rate and Saturation

Definitions

- **Waist of a laser:** The half width of the intensity plot where the intensity is 13.5% of the maximum intensity, see Fig. 3.7. We use the variable w to represent waist.
- **Intensity:** The intensity of a laser beam is the power of the laser divided by the cross sectional area of the laser beam, denoted as $I = P/A$.

If you shine a laser on the wall, you will see something that appears to be a circle. However, it isn't actually a circle. If we plotted the intensity of the laser, it would look like Fig. 3.7a. The waist of a laser beam is defined in Fig. 3.7b. It is a bit

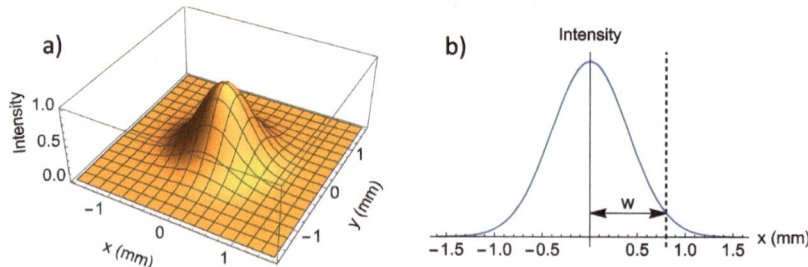

Fig. 3.7 (**a**) The intensity of a laser beam on the wall. (**b**) The intensity along any one of the axes. The waist is the half width of the intensity plot when the intensity is 13.5% of the maximum intensity

---

[2] A careful reader might notice the units for lifetime look like $\frac{s}{rad}$. You can think of radians as simply a placeholder to remind us to use radians (2π) instead of degrees (360°) in our math when calculating frequency. We do not keep the radian unit for lifetime. That unit for lifetime is seconds (or μs or ns).

confusing since most people think of the waist of a laser as the diameter, but it is more similar to a radius. The intensity profile is described by a Gaussian function.[3] The area of a laser beam is:

$$A = \frac{1}{2}\pi w^2 \tag{3.9}$$

This is, of course, for a perfect laser beam. In real life, the intensity profile is not a perfect Gaussian function and the waist in the x direction and y directions could be different. In the latter case, the cross sectional area of a laser is $\frac{1}{2}\pi w_x w_y$.

Guiding Question: We are interested in the following question: Given a laser and a particular transition in an element, how many photons per second will that atom absorb (and re-emit)? This value is known as the **scattering rate**. Take a few moments to think about this question before continuing.

Intensity Versus Power In the lab, we measure laser power. However, we actually care about laser intensity. Why? (This is a homework problem as well).

How many photons an atom scatters depends upon a few different ideas: (1) how far the laser frequency is from the resonance frequency *relative to* the natural linewidth, (2) how long an atom spends in an excited state, and (3) how a transition reacts to photons from a laser whose frequency matches the resonance frequency. Let's unpack all of these ideas.

(1) We now know that the shape of the absorption bump has the mathematical form of a Lorenztian function. For spectroscopists, the natural linewidth really matters. Suppose an atom has a natural linewidth of $\gamma = 100\,\text{MHz}$ and the laser is 25 MHz below the resonance frequency, see the left spectral profile in Fig. 3.8. For this transition, the atom will "scatter" (i.e., absorb and re-emit) many photons from the laser. The right spectral profile in Fig. 3.8 corresponds to a transition that has a natural linewidth of $\gamma = 1\,\text{MHz}$. For this narrower transition, the atom will not scatter many photons at all even though the laser frequency is still set to 25 MHz below the resonance frequency.

To quantify how far the laser frequency is from the resonance frequency, we define a new parameter called **detuning**. Detuning, which is represented by the lowercase Greek letter delta δ in normal frequency units and capital delta $\Delta = 2\pi\delta$ for angular frequency units, gives us this information. Mathematically, detuning is $\delta = f - f_r$, where f is the frequency of the laser. Notice that if $f < f_r$, then $\delta < 0$. In the lab, we call this "red detuning," which will

---

[3] A Gaussian function is very similar to a Lorentzian but it is slightly different. We will explore Gaussian functions more in Chap. 4.4.

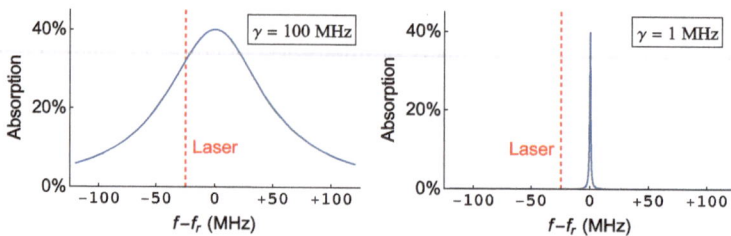

Fig. 3.8 Two examples for how light from a laser interacts with two different transitions. The left transition has a large natural linewidth while the right transition has a small natural linewidth. The laser frequency (red dashed line) is 25 MHz below the resonance frequency for both transitions

make more sense after we discuss the Doppler effect in Chap. 4. Likewise, if $f > f_r$, then $\delta > 0$, which we call "blue detuning." In our thought experiment, the atoms will absorb the largest number of photons when $\delta = 0$ (the laser frequency exactly matches the resonance frequency). In the absence of power broadening (i.e., when the laser power is kept low), the atoms will absorb half as many photons when $\delta = \gamma/2$ or $\delta = -\gamma/2$ compared to when $\delta = 0$.

As discussed above and displayed in Fig. 3.8, detuning is important, but so is the natural linewidth. What we really want to know is how far the laser frequency is from the resonance frequency *relative to* the natural linewidth. For example, if we set the laser frequency such that $\delta = -\gamma/2$, the laser will lose half as many photons compared to when $\delta = 0$. This is true for any transition. The quantity that matters for the scattering rate is the *ratio* of detuning to natural linewidth. So, we expect δ/γ to show up in the relevant scattering rate equation.

(2) The scattering rate tells us how many photons are absorbed (and re-emitted) by an atom. From point (1), we know that detuning matters. The average time an atom spends in the excited state also matters. If an atom, on average, spends only 1 ns in the excited state, it decays very quickly, freeing itself up to be excited again. Contrast that with an atom that spends, on average, 2 seconds in the excited state. That atom will have a very small scattering rate. From Eq. 3.8, we know that the average lifetime is related to the natural linewidth. A large natural linewidth means a short lifetime. Therefore, a large natural linewidth results in a large scattering rate.

(3) Let's consider two transitions with different resonance frequencies. Transition #1 has a resonance frequency f_{r1} and transition #2 has a resonance frequency f_{r2}. We will assume that $f_{r1} > f_{r2}$ and that both transitions have the same natural linewidth. We also have two lasers with the same intensity. The frequency of laser #1 is set to f_{r1} and is sent through a vapor cell with atoms that have transition #1. The frequency of laser #2 is set to f_{r2} and is sent through a vapor cell with atoms that have transition #2. In other words, the only difference between the two experiments is that transitions have different

resonance frequencies. Will both transitions scatter (i.e., absorb and re-emit) the same number of photons per second, or will the scattering rates be different?

Surprisingly, the answer is that the scattering rates are different for the two transitions! The reason is that the cross section (or "size") of the photons is different for the two transitions. Roughly, the cross section of a photon is λ^2. If a transition has a smaller resonance frequency, the laser light has a longer wavelength and photons with a bigger cross section. That means the photons in the laser are "big" and more likely to "hit" the atom and cause the transition. If a photon has a small cross section (high frequency, short wavelength), it is less likely to hit the atom and cause the transition. For the example above, the photons from the laser with $f = f_{r1}$ have a smaller cross section than the photons from the laser with $f = f_{r2}$. If we want the same number of scattered photons from the two transitions, the atom with f_{r1} needs to be exposed to more photons than the transition with f_{r2}. In other words, to achieve the same scattering rate for the two transitions, the transition with f_{r1} requires a higher intensity than the transition with f_{r2}.

Summary

(1) Detuning *relative to* the natural linewidth matters. $\delta = 0$ should have the largest scattering rate.
(2) The natural linewidth (or lifetime of an excited state) matters. Large γ should have a large scattering rate.
(3) Photon cross section matters. A large frequency, or short wavelength, has a smaller scattering rate.

To quantify points (2) and (3), we introduce a parameter known as the **saturation intensity** I_s. The saturation intensity contains all the information about a particular transition that helps us understand how easily an atom interacts with photons in a laser beam whose frequency matches the resonance frequency of a transition (we will quantify this statement soon). Suppose we send a laser beam whose frequency matches the resonance frequency for some transition through a sample of atoms. If an atom absorbs a photon, it will spend some amount of time in the excited state before decaying back to the ground state, where it is free to absorb another photon. The saturation intensity is the laser intensity for an on-resonance laser ($f = f_r$) such that 25% of the atoms are in the excited state at any given time. A transition with a small saturation intensity means that we only need a small laser intensity to have 25% of the atoms in the excited state. A transition with a large saturation intensity means we need a large laser intensity to make that happen.

The ratio of the intensity of light to the saturation intensity is called the **saturation parameter** $s = I/I_s$. We like to use s because, like δ/γ, it means the same thing for every transition. Saying $s = 1$ means that we set the laser intensity equal to the saturation intensity. Some transitions might have a high saturation intensity, like a transition in the beryllium atom that has $I_s = 885\,\mathrm{mW/cm}^2$, while other transitions have a low saturation intensity, like a transition in the cesium atom that has $I_s = 0.40\,\mathrm{mW/cm^2}$. For that transition in the beryllium atom, we would need 885 mW of power for a laser with cross sectional area $A = 1\,\mathrm{cm^2}$ to have $s = 1$, which means 25% of the atoms are in the excited state. 885 mW is a lot of laser power! For the transition in the cesium atom, we only need 0.40 mW of power to have 25% of atoms in the excited state. The formula for the saturation intensity is:[4]

$$I_s = \frac{2\pi^2}{3}\frac{hc\gamma}{\lambda^3}.$$ (3.10)

Notice that the saturation intensity is proportional to γ. This is point (2). Also notice that there is a λ^3 in the denominator. One of those λ terms is grouped with hc in the numerator; hc/λ represents the energy of the on-resonance photon. The remaining λ^2 comes from the cross section of a photon. The saturation intensity is a property of a transition! It is whatever it is, and it cannot be changed.

Assessing the Scattering Rate Formula Before We Even Write It Down
Before we write down the formula that models the scatter rate, decide if δ/γ should be in the numerator or the denominator? In other words, would $\delta/\gamma = 100$ result in more or less photons scattered compared to $\delta/\gamma = 0$? What about the saturation parameter s? Should that parameter be in the numerator or denominator?

The scattering rate, which is derived using quantum mechanics, is how many photons per second an atom will absorb (and re-emit). The scattering rate, $r_\Gamma(\Delta, s)$ using angular frequency variables and $r_\gamma(\delta, s)$ using normal frequency variables, is:

$$r_\Gamma(\Delta, s) = \frac{\Gamma}{2}\frac{s}{1+s+4(\frac{\Delta}{\Gamma})^2}$$

$$r_\gamma(\delta, s) = \pi\gamma\frac{s}{1+s+4(\frac{\delta}{\gamma})^2}.$$ (3.11)

The equation using normal frequency units is more practical, but you will rarely see that formula written anywhere. Almost every atomic physics textbook will use the scattering rate formula that uses angular frequency variables. Notice that the ratio δ/γ is in the denominator. If that number gets big, the scattering rate decreases. The saturation parameter is in the numerator *and* the denominator, which you might not have guessed. It makes sense for it to be in the numerator because if I increased the

---

[4] For completeness, this is the formula for doing spectroscopy with linearly polarized laser light. Experiments with circular polarized light have a slightly different formula.

laser intensity there are more photons for the atom to interact with making it more likely to absorb and emit a photon. But what about the extra s in the denominator?

The first thing to notice is that if the saturation parameter is small, than $1+s \approx 1$, and s would only be in the numerator. The extra s in the denominator comes from the fact an atom will always spend some amount of time in the excited state. If we got rid of the s in the denominator, the number of photons scattered per second (i.e., the scattering rate) is linear with laser power. That means if we increase the laser power by some factor, we increase the scattering rate by the same factor. However, the atom spends, on average, time $\tau = 1/(2\pi\gamma)$ in the excited state. If the atom is already in the excited state, it can't absorb a photon. If all the atoms were in the excited state, there are no atoms left in the ground state to absorb any photons.

This is important, so let's explore it a little more. Suppose an atom always spends a time $\tau = 10\,\text{ns}$ in the excited state before decaying back to the ground state.[5] In the most extreme case of super high laser intensity, an atom would be immediately re-excited back to the excited state before having to wait another 10 ns to decay. In this most extreme case, the atom can only absorb 1 photon every 10 ns. That means, at most, an atom can absorb $\frac{1\,\text{photon}}{10\,\text{ns}} = \frac{1\,\text{photon}}{10\times 10^{-9}\,\text{s}} = 10^8$ photons every second. Without the s in the denominator, the scattering rate would increase without bound as the power increases. However, with the s in the denominator, the on-resonance scattering rate will "saturate" at $\pi\gamma$.[6]

Summary of Formulas

$$r_\gamma(\delta, s) = \pi\gamma \frac{s}{1+s+\frac{4\delta^2}{\gamma^2}}$$

$$I = P/A$$
$$A = \tfrac{1}{2}\pi w^2 \text{ (for a laser beam)}$$
$$s = \frac{I}{I_s}$$ (3.12)
$$\delta = f - f_r$$
$$I_s = \frac{2\pi^2}{3}\frac{hc\gamma}{\lambda^3}$$

One Final Thing: The fraction of atoms in an excited state was one of the key concepts we used to explore the scattering rate. So, it is no surprise that scattering rate also tells us what fraction of atoms are in the excited state. The fraction of atoms in the excited state is $r_\Gamma(\Delta, s)/\Gamma$ (use the angular frequency formulas for calculating the excited state fraction).

---

[5] Remember that an atom actually decays probabilistically with a characteristic time τ. This is just a thought experiment to understand the concept of saturation.

[6] A careful reader might notice the maximum scattering rate is only half as big as our thought experiment predicted. The idea of the thought experiment is correct, but we are ignoring an effect known as coherent state transfer, which includes stimulated absorption and stimulated emission. This is a massive, complicated topic, and one that is beyond the scope of this book. In fact, most quantum mechanics classes don't get to this idea until the very end of the semester, if they get to it at all. Including this extra physics reduces the maximum scattering rate by a factor of 2.

Let's do a quick assessment. If $s = 1$ (or $I = I_s$) and $\Delta = 0$, the excited state fraction should be 25%:

$$\frac{r_\Gamma(0, 1)}{\Gamma} = \frac{1}{2}\frac{1}{1 + 1 + \frac{4(0)^2}{\Gamma^2}} = \frac{1}{4} = 0.25$$

3.5   Power Broadening

The scattering rate tells us how many photons per second an atom takes from the laser. A large scattering rate must correspond to a larger amplitude spectral feature in an absorption plot. In fact, atoms scattering photons is the only way to produce a spectral feature. If that is the case, shouldn't the scattering rate be a Lorentzian function (Eq. 3.1) like our spectral features? It looks close, but there is that extra s in the denominator of the scattering rate that is not in Eq. 3.1. While it might not look like it, the scattering rate is a Lorentzian function. We just need to do a little algebra to convert the scattering rate into a new form. The algebraic step is to factor out $1 + s$ from the denominator:

$$r_\gamma(\delta, s) = \pi\gamma\frac{s}{1 + s + \frac{4\delta^2}{\gamma^2}} = \left(\frac{\pi\gamma}{1 + s}\right)\frac{s}{1 + \frac{4\delta^2}{\gamma^2(1+s)}}. \tag{3.13}$$

We are going to define a new parameter $\gamma_s = \gamma\sqrt{1 + s}$, known as the power broadened linewidth. With this definition, we can write $\frac{4\delta^2}{\gamma^2(1+s)}$ as $\frac{4\delta^2}{\gamma_s^2}$. To help us assess this formula, we will also switch the positions of s and $\pi\gamma$ in the numerator. With those changes, we have:

$$r_\gamma(\delta, s) = \left(\frac{s}{1 + s}\right)\frac{\pi\gamma}{1 + \frac{4\delta^2}{\gamma_s^2}}. \tag{3.14}$$

This is now a Lorentzian function with amplitude $\frac{s}{1+s}\pi\gamma$ and FWHM of $\gamma_s = \gamma\sqrt{1 + s}$. As we increase the saturation parameter s (i.e., increase the laser intensity), the FWHM of the scattering rate becomes larger by a factor of $\sqrt{1 + s}$. Therefore, the width of a spectral feature increases by the same amount. Also notice that when $s \to 0$, the FWHM reaches its minimum value of the natural linewidth.

Next, let's analyse the amplitude. As the saturation parameter gets larger and larger, $1 + s \approx s$, so $\frac{s}{1+s} \to 1$. The amplitude saturates! As the laser power increases, the amplitude approaches the maximum scattering rate of $\pi\gamma$. However, the FWHM never saturates; it continues to broaden, as shown in Fig. 3.9.

Finally, let's put some numbers in to start getting comfortable with real scattering rate numbers. For the example in Fig. 3.9, I used the natural linewidth for a transition in the cesium atom near 852 nm. The natural linewidth is about $\gamma = 5.22$ MHz. The on-resonance ($\delta = 0$) scattering rate is 8.2 million photons absorbed (and re-emitted) per second for $s = 1$; 14.9 million photons per second for $s = 10$;

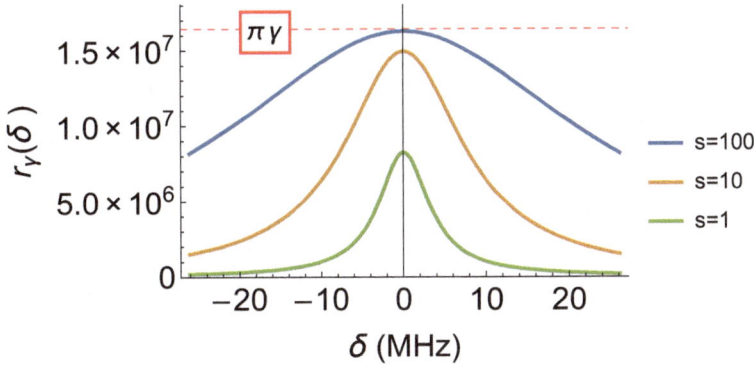

Fig. 3.9 The scattering rate for a transition with $\gamma = 5.22\,\mathrm{MHz}$ as a function of detuning for different saturation parameters. The red dashed line is the maximum possible scattering rate

16.2 million photons per second $s = 100$. For a typical transition, scattering hundreds of thousands to millions of photons per second is not unusual. Notice that when s is small, the scattering rate at $\delta = -20\,\mathrm{MHz}$ is almost 0; mathematically, it is about 270,000 photons/sec for $s = 1$. However, the scattering rate for $s = 100$ remains quite sizable.

For completeness, there are other factors that can broaden the width of a transition including temperature (this is the topic of Chap. 4) and pressure. We do not cover pressure broadening, also known as collisional broadening, in this book.

3.6 Example

A lot has happened in this chapter, but two of the important concepts are that (1) the scattering rate tells us how many photons per second an atom absorbs (and emits) from the laser and (2) the cumulative effect of all the atoms taking photons results in the absorption profile. Let's solidify these concepts with an example. Throughout this example, we will also introduce some commonly used language in atomic physics. A lot of the language will be explored in more detail as we work our way through this book.

A particular type of barium atom[7] called barium-135 has one ground state and three closely spaced excited states, see Fig. 3.10. The spacing between the excited states is called hyperfine splitting, which will be explored in detail in Chap. 9. The reason there are three excited hyperfine states is because the nucleus has angular momentum, a concept we will begin exploring in Chap. 7. If the nucleus did not

---

[7] All barium atoms have 56 protons in the nucleus. Neutron number can vary from about 58–97! We call a particular barium atom, for example an atom with 56 protons and 79 neutrons, an isotope of barium. Isotopes are explored in Chap. 10.

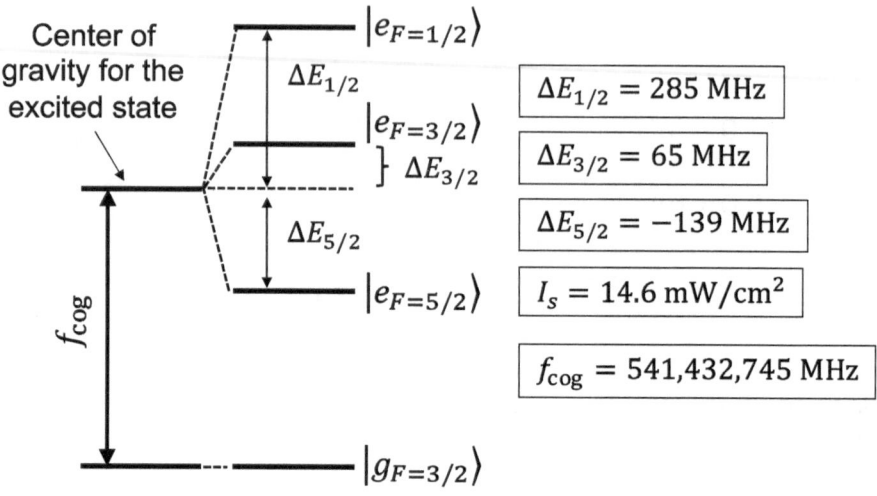

Fig. 3.10 A simplified Grotrian diagram for a transition in the atom known as barium-135. The energy spacings are not to scale. The energy spacings between the excited hyperfine levels are calculated from the results of Baird et al.[1]. The center of gravity frequency is extracted using numbers from both Baird et al. [1] and Karlsson et al. [2]. The uncertainty in the center of gravity frequency is about 30 MHz. [2] While that isn't a terrible uncertainty, modern day spectroscopic methods can do better!

have angular momentum, there would be only one excited state.[8] Again, we will learn the physics behind hyperfine splitting starting in Chap. 7. Before we do an example, we need a few more definitions.

Definitions

- **Center of gravity:** The energy of a state if the nucleus had no angular momentum.
- **Hyperfine splitting:** When the nucleus has angular momentum, the single energy level at the center of gravity splits into multiple energy states. Each of these states is going to shift in energy a small amount compared to the center of gravity energy.
- **Bra-ket notation:** Atomic physicists sometimes use "bra-ket" notation when working with states in an atom. A "ket" is a way to represent a particular state and it looks like this: |put state label here⟩. A "bra", which is not used in this book but you will use it a lot if you take quantum mechanics, looks like this: ⟨put state label here|. Bra-ket notation is also referred to as Dirac notation, named after the English mathematical and theoretical physicist Paul Dirac.

---

[8] There are atoms with ground state hyperfine splitting; barium-135 is just not one of those atoms.

Note
Some learners may use only Part 1 of this book. I wanted to introduce you to bra-ket notation so that you have seen it at least once before you take quantum mechanics. We are going to use it in this example and in Problem 3.9, but it will not be used again until Part 2. You are, of course, welcome to use bra-ket notation if you want, but it is not necessary. You may also see bra-ket notation if you take a linear algebra class.

As seen in Fig. 3.10, we are going to label the ground state $|g_{F=3/2}\rangle$ and the three excited states $|e_{F=1/2}\rangle$, $|e_{F=3/2}\rangle$, and $|e_{F=5/2}\rangle$. F is called a quantum number, and it is always a positive integer, a half integer, or zero; quantum numbers are explored starting in Chap. 6. For now, these quantum numbers are just being used to label our states. Notice there is a center of gravity frequency, f_{cog}, that tells us the energy difference between the center of gravity for the two states. This tells us that our transitions are all around 541.4 THz, or 553.7 nm. Each hyperfine level is shifted from their center of gravity state by a small amount. I want to emphasize that if a nucleus has angular momentum, the center of gravity states do not exist in real life! The hyperfine states are the actual states. However, we can learn a lot of physics by determining how the hyperfine levels shift from the center of gravity, which is explored in Chaps. 9 and 10.

With that nice long intro, let's get into the actual example. We have four states: one ground state and three excited states. We are going to assume that our atoms are all at rest and send a laser beam through the sample, see Fig. 3.1. All of the atoms will start in the ground state $|g_{F=1/2}\rangle$. An electron can be excited to a higher energy level as long as the following rule, derived from quantum mechanics, is satisfied: A transition can change F by -1, 0, or 1. *Memorize this rule!* For completeness, there is one exception to this rule. If the ground state has $F = 0$, that electron cannot be excited to an $F = 0$ excited state.[9]

The Rule

$$\Delta F = -1,\ 0,\ +1 \tag{3.15}$$
$$F = 0 \nrightarrow F = 0$$

In this example, an electron in the barium ground state can be excited to any of the excited states. However, if there were an excited state with quantum number $F = 7/2$, an electron in the ground state could not be excited to this state because $\Delta F = 2$.

The natural linewidth of all three transitions is about $\gamma = 19$ MHz with a saturation intensity of $I_s = 14.6$ mW/cm^2. Let's set our laser intensity to $I =$

---

[9] We only need this rule for the moment. A full list of the rules that need to be satisfied for an electron to transition between two atomic states is given in Appendix C.

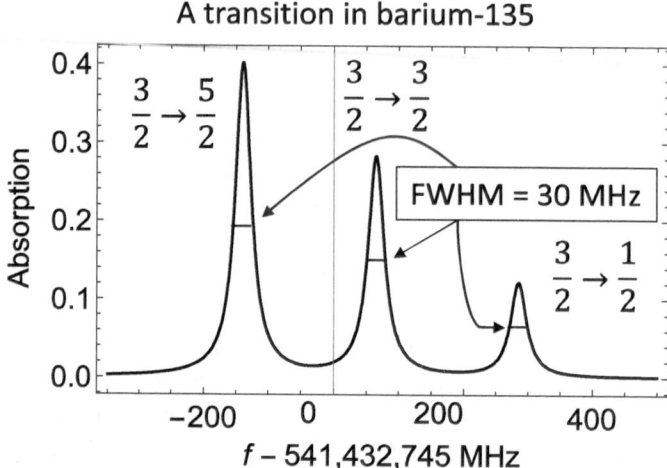

Fig. 3.11 A simulated absorption plot for a transition in barium-135. In this example, we are assuming that all the atoms are at rest and the experimental setup is shown in Fig. 3.1. The horizontal axis is the laser frequency with respect to the center of gravity frequency

$21.9 \, \text{mW/cm}^2$, corresponding to a saturation parameter of $s = 1.5$, and predict an absorption plot. In this example, we are interested in the frequency of each spectral feature, so we are, for now, going to ignore the amplitudes of the spectral features. We will discuss how to calculate the amplitudes in Chap. 9.5.

First, we calculate the resonance frequency for each transition:

$$
\begin{aligned}
\left|g_{F=3/2}\right\rangle \rightarrow \left|e_{F=1/2}\right\rangle : \quad & f_r = f_{\text{cog}} - 139 \, \text{MHz} = 541,432,865 \, \text{MHz} \\
\left|g_{F=3/2}\right\rangle \rightarrow \left|e_{F=3/2}\right\rangle : \quad & f_r = f_{\text{cog}} + 65 \, \text{MHz} = 541,433,069 \, \text{MHz} \\
\left|g_{F=3/2}\right\rangle \rightarrow \left|e_{F=5/2}\right\rangle : \quad & f_r = f_{\text{cog}} + 285 \, \text{MHz} = 541,433,289 \, \text{MHz}
\end{aligned}
$$

These are the center frequencies of each spectral feature, and each feature has a Lorentzian lineshape with a width of $\gamma\sqrt{1+s} = (19 \, \text{MHz})\sqrt{1+1.5} = 30 \, \text{MHz}$. Figure 3.11 shows the simulated results. The horizontal axis represents the laser frequency relative to the center of gravity frequency, $f_{\text{cog}} = 541,432,745 \, \text{MHz}$.

Problems

3.1 Go through the chapter and write down all the new fundamental constants. Don't forget units.

3.2 For each of the formulas in Eqs. 3.12 and 3.15, write a brief description of what each equation means.

3.3 In the lab, we measure laser power. However, we actually care about laser intensity. Why?

3.4 This problem explores the scattering rate, Eq. 3.11.

(a) What is the on-resonance ($\delta = 0$) scattering rate?
(b) At what laser detuning would the scattering rate be half of the on-resonance scattering rate?
(c) Show that in the low power limit (i.e., the limit where s is very small), the answer for part (b) is $\delta = \pm\gamma/2$.
(d) In the low power limit, what would δ be such that an atom absorbs 1/100 as many photons compared to the on-resonance case?

3.5 A transition in the Europium atom has a natural linewidth of $\gamma = 25.5\,\mathrm{MHz}$. The wavelength of light at the resonance frequency is $\lambda = 466.188\,\mathrm{nm}$. Calculate the saturation intensity in units of $\mathrm{mW/cm^2}$ and $\mathrm{mW/mm^2}$.
Hint: 1 W=1 J/s

3.6

(a) For the transition in Problem 3.5, calculate the on resonance scattering rate ($\delta = 0$) for a saturation parameter of 0.1, 1, 5, 10, and 100.
(b) Find the excited state fraction for each of the above saturation parameters.

3.7 Show that the maximum excited state fraction is 50%.

3.8 If a laser beam has a waist of $w = 1\,\mathrm{mm}$, what power should we set the laser in order to get saturation parameters of 0.1, 1, 5, 10, and 100? Assume the saturation intensity is $I_s = 1.2\,\mathrm{mW/mm^2}$.

3.9 (Graphing Problem) Plot the scattering rate versus detuning for a transition with $\gamma = 10\,\mathrm{MHz}$ for different saturation parameters of 0.1, 1, 5, 10, and 100.

3.10 (Rubidium-80 Spectroscopy)
Rubidium-80 is an atom with 37 protons (all isotopes of rubidium have 37 protons) and 43 neutrons. It is an unstable atom that undergoes radioactive decay to krypton-80. Radioactive decay is discussed in Chap. 10. For this problem, we want to predict an absorption plot assuming all the atoms are at rest.

The transition we are interested in is shown in Fig. 3.12. There are two ground state hyperfine levels labeled $\left|g_{F=1/2}\right\rangle$ and $\left|g_{F=3/2}\right\rangle$. The excited state has three hyperfine levels labeled $\left|e_{F=1/2}\right\rangle$, $\left|e_{F=3/2}\right\rangle$, and $\left|e_{F=5/2}\right\rangle$. The natural linewidth of this transition is $\gamma = 5\,\mathrm{MHz}$ and we set the laser intensity such that $s = 3$.

(a) Using the rule shown in Eq. 3.15, find the resonance frequency for all possible transitions (there are five of them).

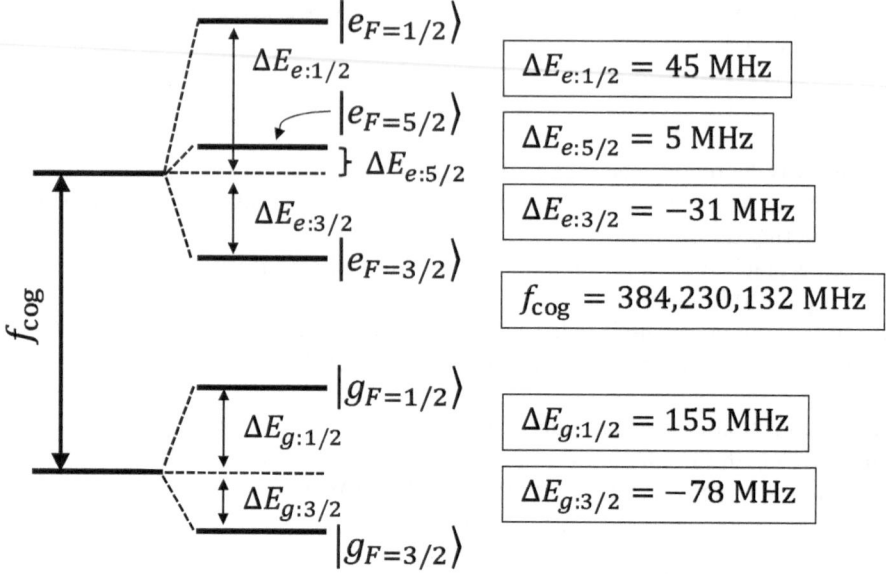

Fig. 3.12 A simplified Grotrian diagram for a transition in the atom known as rubidium-80. The energy spacings are not to scale. The energy spacings are taken from the work of Thibault et al. [3]

(b) What is the width of each spectral feature?

(c) Make an absorption plot with respect to the center of gravity frequency; see Fig. 3.11 for an example. Make all the amplitudes the same. There is a formula to calculate the relative amplitudes, but we won't talk about that until Chap. 9.

References

1. Baird, P.E.G., Brambley, R.J., Burnett, K., Stacey, D.N., Warrington, D.M., Woodgate, G.K.: Optical isotope shifts and hyperfine structure in λ553.5 nm of barium, Proc. R. Soc. Lond. A365567–365582 (1979). http://doi.org/10.1098/rspa.1979.0035

2. Karlsson, H., Litzén, U.: Revised Ba I and Ba II wavelengths and energy levels derived by fourier transform spectroscopy. Phys. Scripta **60**, 321 (1999). https://doi.org/10.1238/Physica. Regular.060a00321

3. Thibault, C., Touchard, F., Büttgenbach, S., Klapisch, R., de Saint Simon, M., Duong, H.T., Jacquinot, P., Juncar, P., Liberman, S., Pillet, P., Pinard, J., Vialle, J.L., Pesnelle, A., Huber, G.: Hyperfine structure and isotope shift of the D$_2$ line of 76–98Rb and some of their isomers. Phys. Rev. C **23**, 2720 (1981). https://doi.org/10.1103/PhysRevC.23.2720

Atoms in Motion

4

Abstract

In this chapter, we explore how motion can affect the perceived frequency of waves, with significant implications for spectroscopy. We cover the broadening and shifting of atomic line frequencies due to atomic movement and investigate the roles of velocity and temperature in these phenomena. Key topics include the Doppler effect, Doppler broadening, the Maxwell-Boltzmann velocity distribution, and the Equipartition Theorem. Additionally, the chapter discusses the application of the Doppler effect in astronomy.

Learning Goals

By the end of this chapter, you should be able to understand:

- the Doppler effect.
- Doppler broadening and how temperature affects the spectral features in absorption and transmission plots.
- the Maxwell-Boltzmann velocity distribution.
- how the temperature of a vapor cell is related to the average speed of an atom in that cell.
- the application of the Doppler effect in astronomy.

4.1   The Doppler Effect

The Doppler effect is likely a phenomenon you have encountered before. When an ambulance, police car, or racecar travels past you, the sound you hear changes pitch. This happens because the motion of the vehicle compresses or extends the

© The Author(s) 2025
W. Raven, *Atomic Physics for Everyone*,
https://doi.org/10.1007/978-3-031-69507-0_4

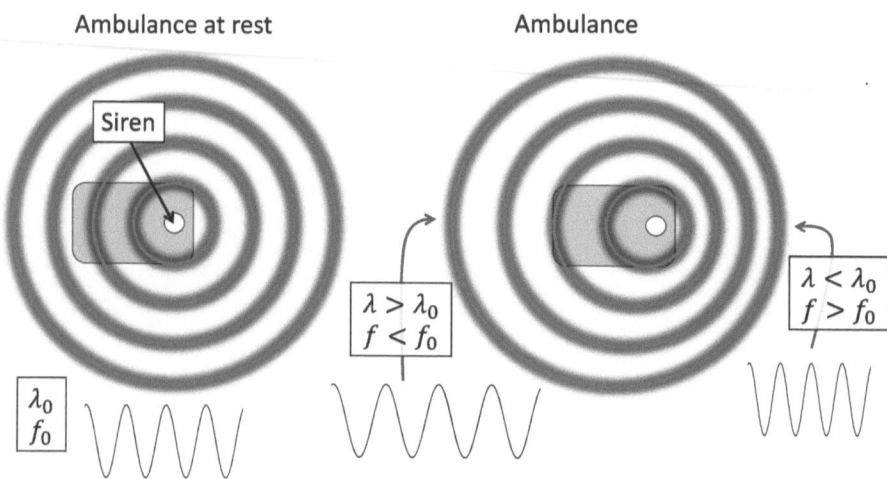

Fig. 4.1 Left: An ambulance at rest emitting a sound wave from its siren. The wavelength λ_0 and frequency f_0 of the sound wave is the same in all directions. Right: Now the ambulance is moving to the right. The sound wave in front of the ambulance is compressed, which means the perceived wavelength is smaller and the perceived frequency is larger (higher pitch). The sound wave behind the ambulance is expanded, which means the perceived wavelength is larger and the perceived frequency is smaller (lower pitch)

sound waves. Figure 4.1 shows the sound waves emitted by a stationary ambulance (left) and a moving ambulance (right). Let's focus on the stationary ambulance. Imagine that you are standing in front of or behind the ambulance. The wavelength of the sound wave that hits your ear is the same for both scenarios, so you would hear the same pitch independent of where you are standing. Now, imagine the ambulance is moving. If you were standing in front of the ambulance (OK, maybe a bit to the side ... we don't want you to get hit, even in a thought experiment), the wavelength of the sound wave that reaches your ear is shorter compared to the stationary ambulance. If you were standing behind the moving ambulance, the wavelength is longer compared to the stationary ambulance. The formula that relates the frequency (pitch) that you hear to the wavelength should look really familiar. It is $v_s = f\lambda$, where v_s is the speed of sound in air (replace v_s with c and you have Eq. 1.1 from p. 19). The apparent shift in frequency due to an object moving is known as the Doppler effect, named after Austrian physicist and mathematician Christian Doppler. It is a very important concept in spectroscopy.

The Doppler effect occurs for any type of wave. Whether it is a sound wave, a light wave, or a water wave created by a duck swimming in a pond, the relative motion of the object with respect to the observer will change the wavelength, and thus the frequency of the wave. The EMT driving the ambulance hears no change in pitch because they are stationary with respect to the siren. If you yelled positive encouragement at the ambulance as it passed, the driver would hear your pitch change as they passed by you. Likewise, you don't hear your pitch change as the ambulance passes by you. What is important here is that the Doppler effect is something experienced by the observer because the source of the wave is moving with respect to them.

Definition

- **Doppler effect:** An increase or decrease in the frequency of sound, light, or other waves as the source and observer move toward or away from each other.

4.2   Laser Frequency From an Atom's Perspective

What does this have to do with spectroscopy? In Chap. 3, we made an important statement after analyzing Eq. 3.1 on p. 45. It is so important we will repeat it here:

▶   **Important Statement:** A photon has a probability of being absorbed by an atom depending upon the photon's energy. It is most likely to be absorbed if the photon's energy exactly matches the energy difference between the ground and excited state, but there is a non-zero probability of absorption off resonance.
    Specifically, in the absence of any broadening mechanism (like laser power), if the photon's energy is off by $\frac{h\gamma}{2}$ from that resonance energy, the photon is half as likely to be absorbed compared to a photon that has the energy that equals the energy difference between the ground and excited states.

Important Reminder
Frequency, energy, and wavelength are all the same quantity. Each of these parameters is related to the other parameters only by constants.

Imagine a laser beam traveling to the right, as shown in Fig. 4.2. Also imagine there are three atoms: atom 2 is traveling to the left, atom 1 is stationary, and atom 3 is traveling to the right. For this thought experiment, we will assume the speeds of atom 2 and atom 3 are the same, just in opposite directions.

In this experiment, the atom is the observer because it is interacting with the laser light and not producing it. To understand the Doppler effect, it is important to recognize that each atom perceives itself as stationary. Atom 2 would claim that

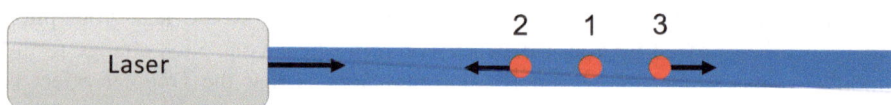

Fig. 4.2 A simple experimental to explore how motion of atoms impacts the interactions between the atoms and laser light

atom 1 is moving to the right and that atom 3 is moving twice as fast as we (as the scientists looking from the outside) would say atom 3 is moving. Both atom 1 and us, as the observing scientists, will agree on the frequency of the laser. Because of the Doppler effect, atoms 2 and 3 will disagree. To make this idea a little clearer, let's say that the laser frequency is 652.0000×10^{12} Hz $= 652.0000$ THz (terahertz) and that this is the resonance frequency for the atom. Both the scientists and atom 1 will agree that the laser frequency is 652.0000 THz; atom 1 will absorb photons from the laser beam. However, atom 2 and atom 3 will disagree with this claim since atom 2 is moving towards the laser and atom 3 is moving away from the laser.

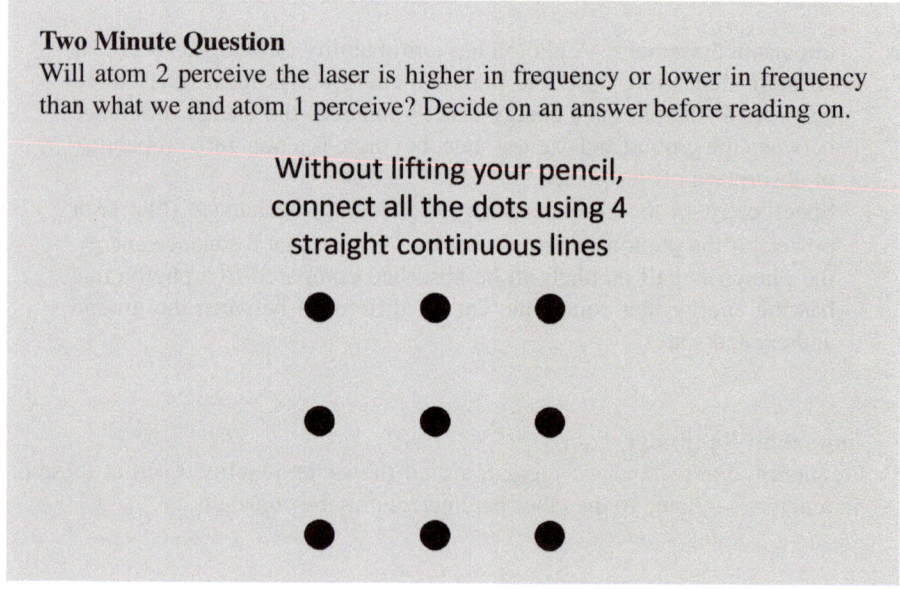

Two Minute Question
Will atom 2 perceive the laser is higher in frequency or lower in frequency than what we and atom 1 perceive? Decide on an answer before reading on.

Answer Atom 2 is moving towards the laser source, so it will perceive the laser frequency as higher than it actually is. Atom 2 will only absorb a photon from the laser if it thinks the laser frequency matches the resonance frequency. Therefore, we, in the observing frame, need to set the laser frequency *smaller* than the resonance frequency (652.0000 THz) so that the actual laser frequency plus the frequency shift due to the Doppler effect results in the resonance frequency in the frame of atom 2. In equation form, this is represented as:

$$f_{\text{atom2}} = f_L + \Delta f_D, \tag{4.1}$$

where f_{atom2} is the laser frequency according to atom 2, f_L is the actual laser frequency (i.e. the frequency measured in the laboratory/stationary frame), and Δf_D is perceived shift in frequency due to the Doppler effect. In this example, $\Delta f_D > 0$ for atom 2, so if we want $f_{\text{atom2}} = f_r$, then we need to set the laser frequency smaller than the resonance frequency such that $f_r = f_L + \Delta f_D$.

Likewise, atom 3 is moving away from the laser source, so it will claim the laser frequency is lower. As a result, the actual laser frequency will have to be higher than the resonance frequency for atom 3 to absorb a photon.

Understanding the Doppler effect is really important in spectroscopy. Inside a gaseous sample of atoms some atoms are moving towards the laser, some are moving away, and some are not moving towards or away from the laser. As we, in the laboratory/stationary frame, change the frequency of the laser from below the resonance frequency to above the resonance frequency, we will find that atom 2 will absorb light at a different frequency than atom 1 or atom 3. Each atom will claim it is absorbing light at precisely the frequency needed to excite it from the ground state to the excited state, and each atom is correct! Atom 2 "sees" a higher frequency than the actual frequency of the laser. When atom 2 "sees" the correct frequency, it will absorb light. For us in the laboratory frame, the laser frequency is too low. This can be confusing, so here is a summary:

- From atom 2's reference frame, the frequency of the laser is just right to excite atom 2 from the ground state to the excited state.
- From the laboratory reference frame, the frequency of the laser is too low.
- The Doppler effect tells us that because atom 2 is traveling towards the laser beam, it will see a higher frequency than what we, as scientists in the laboratory/stationary frame, measure.

An Important Correction I simplified the above description by just a little bit. An atom moves in three dimensions, but only the component of the atom's velocity in the direction toward or away from the laser beam contributes to the Doppler effect. An atom that isn't moving toward or away from a laser can still be moving; it is just moving perpendicular to the laser. A more correct statement is: Inside a gaseous sample of atoms, some atoms have a velocity component pointing towards the laser, some atoms have a velocity component that is pointing away from the laser, and some atoms have no velocity components pointing towards or away from the laser.

We represent that velocity component with the parameter $v_{\parallel}$. It is defined to be positive if the atom is traveling away the laser and negative if traveling towards from the laser.

Definitions

- **Doppler shift:** The shift in the frequency of a laser seen by an atom due to the Doppler effect.
- $v_\parallel$: The velocity component of an atom in the direction of the laser beam. $v_\parallel$ is negative if the atom is traveling towards the laser and positive if it is traveling away from the laser.

The frequency of laser light as seen by the atom is given by the formula:

$$f_{\text{atom}} = f_L \left(1 - \frac{v_\parallel}{c} \right) \tag{4.2}$$

where f_{atom} is the frequency of the laser as seen by the atom, f_L is the frequency of the laser measured in the lab (rest) frame, and $v_\parallel$ is the velocity component in the direction of (parallel to) the laser beam. Using $c = f_L \lambda$, the right-hand side of Eq. 4.2 can be written as:

$$f_{\text{atom}} = f_L - \frac{v_\parallel}{\lambda}, \tag{4.3}$$

where λ is the wavelength of light measured in the laboratory frame.

Important

$v_\parallel$ is a weird variable in that the sign of $v_\parallel$ depends on whether the atom is moving towards ($v_\parallel < 0$) or away ($v_\parallel > 0$) from the laser. Messing up the sign of $v_\parallel$ is a very common mistake when using this formula. When in doubt, just remember that an atom moving towards the laser beam sees a higher frequency.

Comparing Eqs. 4.1 and 4.3, we find the formula for the Doppler shift:[1]

$$\Delta f_D = -\frac{v_\parallel}{\lambda}. \tag{4.4}$$

---

[1] The full formula is $\Delta f_D = -\frac{v}{\lambda} \cos\theta$, where v is the speed of the atom and θ is the angle between the laser and the velocity of the atom ($\theta = 0$ for an atom moving in the same direction as the laser and $180° = \pi$ rad for an atom moving in the opposite direction). The component of the velocity in the direction of the laser is $v_\parallel = v \cos\theta$. If you have worked with vectors before, you might recognize these as vector components.

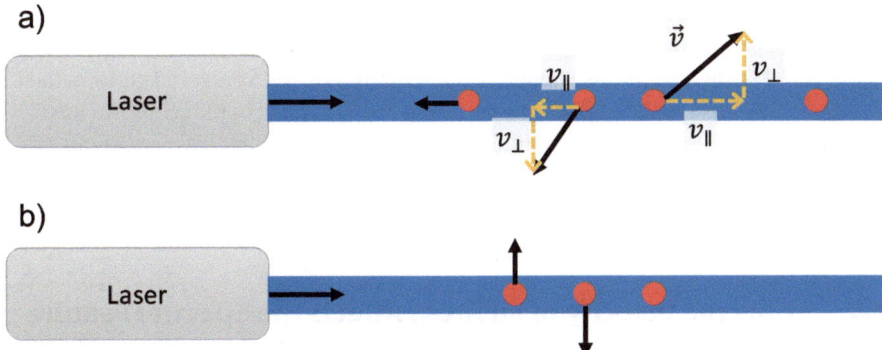

Fig. 4.3 (**a**) Only the component of velocity in the direction of the laser beams results in a Doppler shift. The first two atoms have different velocities, but the same component in the direction of the laser, $v_\parallel$. As such, they will experience the same Doppler shift. The third atom has a velocity component in the opposite direction, so it will have a different Doppler shift. The last atom is completely stationary. (**b**) All three of these atoms have no velocity component in the direction of the laser, so they all have zero Doppler shift

A Bit More About Velocity Components The velocity component in the direction of the laser beam is an important, but sometimes confusing, idea when you first encounter it. So, let's spend a bit more time thinking this idea through using Fig. 4.3. In Fig. 4.3a, the first atom's velocity is pointing directly towards the laser, so $v_\parallel < 0$. For this atom, there is no perpendicular component to the atom, $v_\perp = 0$. If the laser was traveling towards the left, $v_\parallel > 0$ for this atom because the sign of $v_\parallel$ only depends upon if the atom is moving towards or away from the laser beam.

The second atom has both a perpendicular component and a parallel component. Only the parallel component causes the Doppler shift, and the parallel component tells us the atom is moving towards the laser source, so $v_\parallel < 0$. Notice the parallel component for the two first two atoms are the same size and pointing in the same direction. Therefore, they will have the same Doppler shift.

The third atom has a perpendicular component, which we don't care about, and a parallel component pointing away from the laser, so $v_\parallel > 0$. This atom will absorb photons with a different laser frequency than atoms 1 and 2. The last atom is not moving at all. It has no perpendicular or parallel component: $v_\parallel = 0$ and $v_\perp = 0$. Figure 4.3b shows three examples of atoms with $v_\parallel = 0$. Each of these atoms will absorb photons when $f_L = f_r$.

Important
In high precision spectroscopy, we want to extract information from atoms that have $v_\parallel = 0$. The rest of the atoms in our sample make our spectrum

(continued)

less precise. How the Doppler shift changes an absorption plot is going to be explored in the rest of this chapter. Chapter 5 introduces a clever experimental trick known as saturated absorption spectroscopy. This technique allows us to remove the issues brought about from the Doppler shifts that we are about to discuss.

4.3   How the Velocity of an Atom Affects the Spectral Feature

Let's go back to a simplified picture where we have 3 atoms.

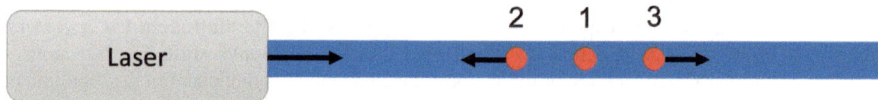

Atom 1 is at rest, atom 2 is traveling towards the laser at speed v, and atom 3 is moving away from the laser at the same speed as atom 2. For this thought experiment, each atom that experiences a Doppler shift will have $|\Delta f_D| = 150\,\text{MHz}$.

What do the transmission and absorption plots look like? Spend a few moments thinking about it, make a prediction, and then read on! Hint: There are only 3 atoms, so the fraction of light lost is very, very small.

The answer is shown in Fig. 4.4. There are three spectral features. Each spectral feature is identical except for a horizontal offset determined by the Doppler shift formula. Since atom 2 is traveling towards the laser beam, it perceives a higher laser frequency compared to what we measure in the lab. Therefore, atom 2 will absorb photons when the laser frequency is below the resonance frequency. The spectral feature from atom 2 is at $\delta = -150\,\text{MHz}$. Likewise, atom 3 is traveling away from the laser, so it is seeing a lower laser frequency compared to what we measure in the lab. Therefore, atom 3 will absorb photons when the laser frequency is above

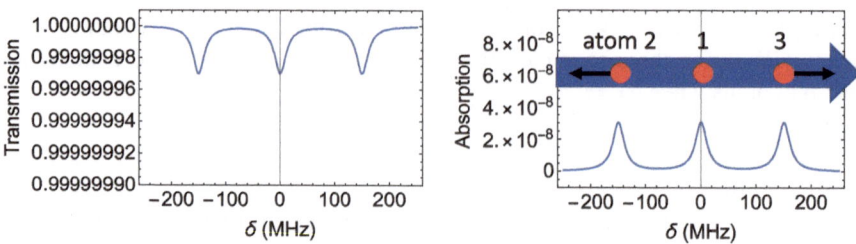

Fig. 4.4 A simulated transmission plot (left) and absorption plot (right) for the three atoms

Table 4.1 A table summarizing the Doppler shifts for the 3 atoms

	Atom 2	Atom 1	Atom 3
Motion	Towards laser	Stationary	Away from laser
The atom sees	Higher frequency light	Actual laser frequency	Lower frequency light
Resonance happens	At lower frequency	At actual frequency	At higher frequency

the resonance frequency. The spectral feature for atom 3 is at $\delta = +150\,\text{MHz}$. This important information is summarized in Table 4.1.

In a real vapor cell, atoms are moving with all sorts of different velocities. Unlike energy levels in an atom, velocity is a continuous variable. In the next two sections, we are going to discuss the distributions of velocities inside a real vapor cell (Sect. 4.4) and use that information to develop what the transmission and absorption plots look like for a vapor cell of atoms at a given temperature (Sect. 4.5).

4.4   The Maxwell-Boltzmann Velocity Distribution

Within a gaseous cloud of atoms, there is a distribution of velocities. This distribution depends on the temperature and mass of the atoms. The distribution of velocities in the direction of the laser beam, which is known as the Maxwell-Boltzmann velocity distribution,[2] is shown by the blue line in Fig. 4.5. The Maxwell-Boltzmann velocity distribution often uses the function $f(v)$ to represent the distribution of velocities. This is not the f we use for frequency. A good rule of thumb is that if it is a function, like $f(v)$, the "f" is probably referring to a distribution. If the "f" is all by itself in a formula, it is probably referring to a frequency.

Fig. 4.5 A cloud of gaseous atoms will have a distribution of velocities given by this graph. This distribution, which is derived from thermodynamics, is known as the Maxwell-Boltzmann velocity distribution

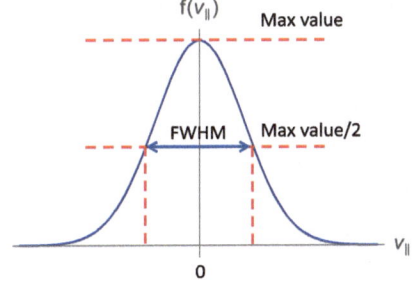

----

[2] Named after the Scottish mathematician James Clerk Maxwell and the Austrian physicist Ludwig Boltzmann.

The full width at half maximum of a Maxwell-Boltzmann velocity distribution, which is a velocity with units m/s, is:

$$\Delta v_{\text{FWHM}} = 2.355\sqrt{\frac{k_B T}{m}}, \qquad (4.5)$$

where $k_B = 1.38 \times 10^{-23}$ J/K is the Boltzmann constant, T is the temperature of the gas (the unit is kelvin), and m is the mass of an atom in the gas (the unit is kilogram). The hotter the gas, the wider the velocity distribution, and thus the larger the average speed of the atoms in the gas. The mass of an atom is in the denominator, so atoms with larger masses have smaller average speeds compared to an equally hot gas of smaller mass atoms.

How to Use Distributions This section isn't really needed to understand the velocity distribution. However, distributions are incredibly important in many areas of science, so I wanted to spend a bit of time talking about how we use them. The velocity distribution represents the fraction of atoms that fall within a particular velocity range. Since an atom must have some velocity, the total area under the curve in Fig. 4.5 is 1. This is equivalent to saying that each atom must have some velocity between $+\infty$ and $-\infty$.

To use a distribution, you ask questions like, "What fraction of atoms have a positive parallel velocity component?", "What fraction of atoms have a parallel velocity component between -2 and 10 m/s?", or more generally, "What fraction of atoms have a parallel velocity component between v_a and v_b?", where v_a and v_b are any velocities we choose.

The answer is the area under the distribution between v_a and v_b. If we want to know the total number of atoms from the sample that have velocity components between those two values, we multiply that fraction by the total number of atoms in the sample. For example, here are three plots with different choices of v_a and v_b:

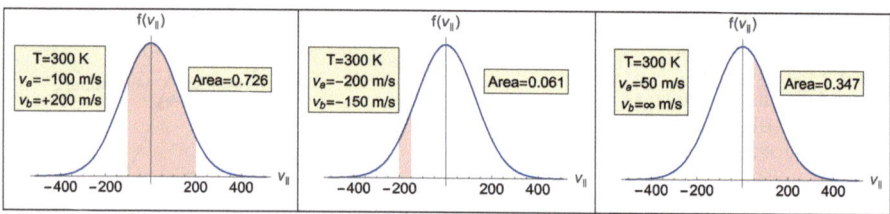

Suppose we have 5000 atoms in our sample. As shown in the first plot, there are $5000 \times 0.726 = 3630$ atoms with parallel velocity components between $v_a = -100$ m/s and $v_b = +200$ m/s. As shown in the second plot, there are $5000 \times 0.061 = 305$ atoms with parallel velocity components between $v_a = -200$ m/s and $v_b = -150$ m/s, and in the third plot there are $5000 \times 0.347 = 1735$ atoms with parallel velocity components larger than 50 m/s.

Fig. 4.6 A comparison between a Gaussian function and a Lorentzian function. Each function has an area under the curve of 1 and a FWHM of 1

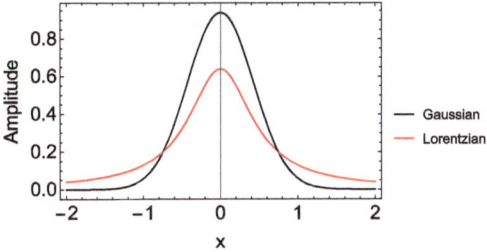

Important Reminder
Due to the Doppler effect, atoms with different velocities will absorb photons at different laser frequencies.

The mathematical function that describes the shape of the Maxwell-Boltzmann velocity distribution is a Gaussian function given by:[3]

$$f_{v_\parallel} = \left(\frac{m}{2\pi k_B T}\right)^{1/2} e^{-\frac{mv_\parallel^2}{2k_B T}} \tag{4.6}$$

A Gaussian looks similar to the Lorentzian function that models the shape of a spectral feature, but the two functions are different. Figure 4.6 shows a plot of both a Gaussian function and a Lorentzian function with the same area and FWHM. Notice the Lorentzian has larger "tails" and is more spread out compared to the Gaussian function. Both functions are very common in physics and math.

Extra Math for Those Who Have Taken Statistics In statistics, Gaussian functions are written as $e^{-\frac{v^2}{2\sigma^2}}$, where σ is called the standard deviation. For the Maxwell-Boltzmann velocity distribution, the standard deviation is $\sqrt{\frac{k_B T}{m}}$. The FWHM of a Gaussian function defined using the standard deviation is:

$$\Delta v_{\text{FWHM}} = 2\sqrt{2\ln(2)}\sigma \approx 2.355\sigma = 2.355\sqrt{\frac{k_B T}{m}} \tag{4.7}$$

---

[3] This is a distribution for 1 dimension since we are only interested in the velocity component for a single direction. In the future, you might encounter a Maxwell-Boltzmann velocity distribution that has a power of 3/2 instead of 1/2 on the expression in front of the Gaussian function. That would be a velocity distribution for all components of velocity, not just the parallel component.

4.5   Transmission and Absorption Plots for Atoms at a Non-Zero Temperature

> **Definitions**
>
> - **Doppler broadening:** The widening of a spectral feature due to a vapor cell having a given temperature.
> - **Doppler profile:** The name given to a spectral feature that is broadened because of temperature.
> - **Doppler width:** The FWHM of a Doppler profile.

In Sects. 4.2 and 4.3, we explored a transmission plot with only three atoms with different velocities. What if we had one hundred thousand atoms? Figure 4.7 shows

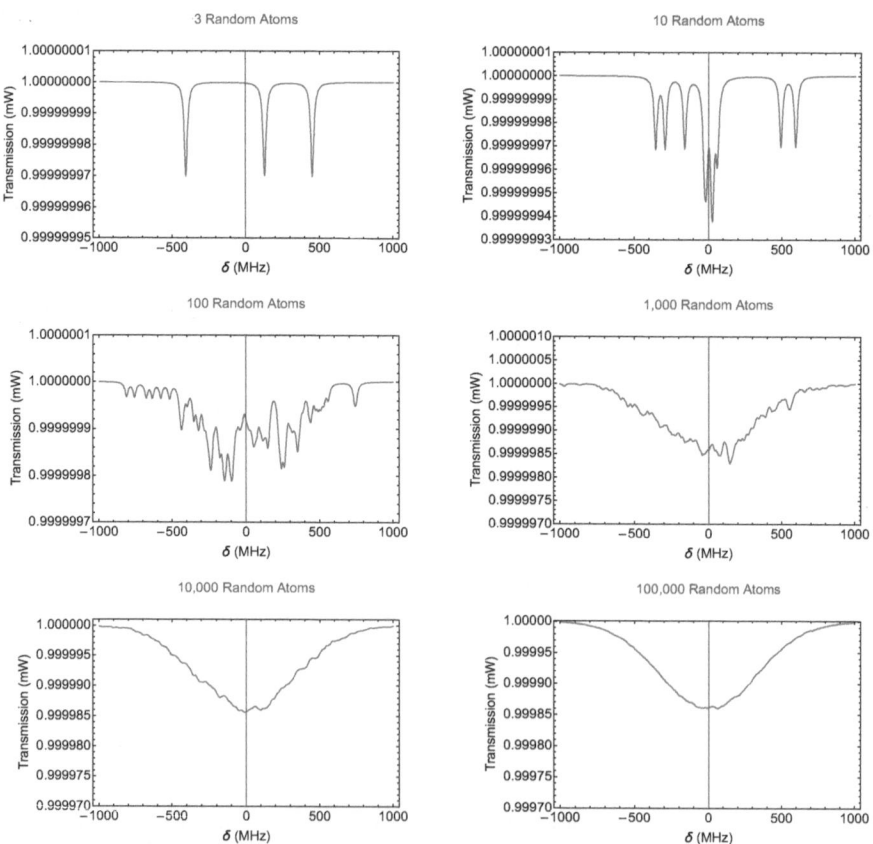

Fig. 4.7 Building a Doppler profile from individual atoms

the results for a simulation of transmission plots as we add more and more atoms to a vapor cell. For this simulation, I assumed that we had a two-level atom with a mass of $m = 2.33 \times 10^{-26}$ kg (this is the mass of a nitrogen atom), a vapor cell temperature of $T = 400$ K, an excitation wavelength of $\lambda = 940$ nm ($f \approx 319$ THz), and a natural linewidth of $\gamma = 5$ MHz. I randomly picked a velocity component using the Maxwell-Boltzmann velocity distribution for each atom I add to the cell.

Note that each transmission plot has a different vertical scale. Individually, a single atom isn't going to absorb a large fraction of a laser's photons. However, the more atoms you have interacting with the light, the more spectral features you have piling up on each other. Ultimately, you get a transmission plot that looks like it has a single feature.

This feature, which is called a **Doppler profile**, is much wider than a spectral feature from a single atom and is pretty close to the same shape as $1 - A f(v_{\parallel})$, where A is some constant and $f(v_{\parallel})$ is the Maxwell-Boltzmann velocity distribution. Notice the center of the Doppler profile is still at the resonance frequency $\delta = 0$.

The width of a Doppler profile can be found from FWHM of the Maxwell-Boltzmann velocity distribution, which is a velocity. We can convert this velocity to a frequency using the Doppler shift formula. The FWHM of a spectral feature broadened by temperature, which is called the Doppler width, is given by the formula:

$$\Delta f_{\text{FWHM}} = \frac{2.355}{\lambda} \sqrt{\frac{k_B T}{m}} \tag{4.8}$$

where Δf_{FWHM} is the Doppler width, which has frequency units.

4.6   The Equipartition Theorem

There is a neat theorem from thermodynamics known as the Equipartition Theorem. Before discussing the Equipartition Theorem, we need to understand **kinetic energy**. Kinetic energy is the energy of movement. Any object with mass m and speed v has kinetic energy:

$$K = \frac{1}{2}mv^2. \tag{4.9}$$

Since kinetic energy is a type of energy, the unit is a joule. Imagine you have 3 atoms in your gas. We will assume that all the atoms have the same mass but different speeds. The average kinetic energy of the atoms in the gas would be:

$$\frac{1}{3}\left(\frac{1}{2}mv_1^2 + \frac{1}{2}mv_2^2 + \frac{1}{2}mv_3^2\right). \tag{4.10}$$

If we had N atoms in our gas, all with the same mass, the average kinetic energy would be:

$$\frac{1}{N}\left(\frac{1}{2}mv_1^2 + \frac{1}{2}mv_2^2 + \dots + \frac{1}{2}mv_N^2\right) = \frac{1}{2}m\left(\frac{v_1^2 + v_2^2 + \dots + v_N^2}{N}\right). \qquad (4.11)$$

That last term that is in parentheses is called the average squared speed. We denote this last term as $\langle v^2 \rangle$. In fact, whenever you see the mathematical expression between two angle brackets, $\langle\ \rangle$, you are being asked to take the average of that property. $\sqrt{\langle v^2 \rangle}$ has a special name which is called the root mean squared speed v_{rms}, which you may have learned about in high school Chemistry. Putting this all together, we find that the average kinetic energy of all the atoms in the gas is:

$$\langle K \rangle = \frac{1}{2}m\langle v^2 \rangle. \qquad (4.12)$$

Reading the Above Equation For a gas composed of atoms with the same mass, the average kinetic energy of the atoms is proportional to the average squared speed.

The Equipartition Theorem tells us that the energy of a gas is equally distributed among all "degrees of freedom." Degrees of freedom indicate the number of ways an atom or molecule can move. Atoms have three degrees of freedom because they can move in three dimensions. Molecules have more degrees of freedom because they can rotate and vibrate, so a molecule has more ways to distribute its energy than an atom. The Equipartition Theorem tells us that each degree of freedom has $\frac{1}{2}k_BT$ of energy. In this book, we are only working with atoms, but, in the future, if you work with molecules, the following formulas will be slightly different.

Imagine you have a vapor cell of atoms at some temperature T. The average kinetic energy of the atoms in the gas is:[4]

$$\frac{1}{2}m\langle v^2 \rangle = \frac{3}{2}k_BT. \qquad (4.13)$$

The 3 on the right hand side represents the three degrees of freedom of an atom $\left(\frac{1}{2}k_BT$ for each degree of freedom$\right)$. This formula is very useful because it directly relates the temperature of a sample of atoms to a characteristic speed of the atoms, specifically $\langle v^2 \rangle$.

---

[4] You can derive this formula from the Maxwell-Boltzmann velocity distribution, but you will need to use calculus.

4.7 Application to Astronomy: Light from the Stars

In our experiment, both we, the scientists, and the laser light source are stationary while the atoms, which act like the observers of the laser light, are moving. The same principles of the Doppler effect apply whether the light source is moving and the observer is stationary, the light source is stationary and the observer is moving, or if both are moving. All that matters is whether the source and observer are moving towards each other or away from each other.

The Doppler effect is a powerful tool in astronomy. Suppose we are using a telescope to collect light from a distant star that is mostly composed of hydrogen gas. That star is emitting light with frequencies corresponding to the difference of the hydrogen energy levels. We now know that if the star is moving towards us, the frequency of light leaving that star will look to us to have a higher frequency than what we would observe if we just had a hydrogen light bulb in our lab. In astronomy, this phenomenon is called blue-shifted light because the light has a higher frequency than we would expect if we measured the spectrum of hydrogen here on earth. If the star is moving away from us, which is far more common in astronomy, the frequency of light that is emitted from the star looks to be lower frequency compared to what we would measure from a source here on earth. This is called red-shifted light.

In summary, if the star is moving towards us, we will see a spectrum that is shifted to higher frequencies compared to what we measure in the lab (blue-shifted; $v < 0$), and we will see a shift to lower frequencies if the star was moving away from us (red-shifted; $v > 0$).

The Doppler effect allows us to calculate the speed of that galaxy. The Doppler formula for a star moving towards (or away) from the earth has a slightly different form than Eq. 4.3:

$$f_{\text{obs}} = \frac{f_{\text{em}}}{1+z}$$
$$z = \frac{v}{c}, \tag{4.14}$$

where f_{obs} is the Doppler shifted frequency measured on earth, f_{em} is the frequency of the light emitted from the star, and v is the speed of the star or galaxy in the direction of earth. Astronomers also use the parameter $z = \frac{v}{c}$ to describe blue-shift light ($z < 0 \rightarrow v < 0$; the star is moving towards the earth) and red-shifted light ($z > 0 \rightarrow v > 0$; the star is moving away the earth). You will have the opportunity to derive this formula in Problem 4.7.

Finally, astronomers like to use wavelength instead of frequency. Writing Eq. 4.14 using wavelength and solving for z gives:

$$z = \frac{\lambda_{\text{obs}}}{\lambda_{\text{em}}} - 1,$$
the formula astronomers use
$$\tag{4.15}$$

Problems

4.1 For each of the following equations, write a brief description of what each equation means.

(a) Equation 4.3
(b) Equation 4.8
(c) Equation 4.13

4.2 Assess Eq. 4.8. The purpose of any assessment is to increase or decrease our confidence in something. Assessments are challenging because we inherently want our calculations to be correct! To combat this bias for assessing a formula, I find it is easiest to write down all the parameters on the right hand side and then try to forget the formula all together. Then you ask yourself the question, "If I increased T, then the Doppler width should get _____ because _____." You need to decide if "larger" or "smaller" goes into the first blank and explain, using a physics reason, why that should happen in the second blank. Next, repeat that process for every parameter. After I think through each parameter, I go check the formula to make sure my statements match the formula.

  If your statement does not match your formula, then either your formula is wrong or your reasoning is wrong. Either way, you now have an opportunity to learn something! But, more importantly, you will understand an equation more after you assess it.

4.3 An atom at rest is excited from the ground state to an excited state by a photon from a laser with frequency $f = 315.11254$ THz.

(a) Suppose the laser is positioned to send photons to the right, and an atom is moving towards the laser with a velocity component of $v_\parallel = -200$ m/s (the minus sign indicates the atom is moving towards from the laser), see atom 2 from Fig. 4.2 on p. 68. What frequency should the laser be for this atom to absorb a photon?
(b) Now the laser is pointed to send photons to the left, so now the atom is moving away from the laser source. What frequency should the laser be for this atom to absorb a photon?

4.4 Explain qualitatively how the motion of an atom affects the energy (frequency) of a photon it will absorb compared to an atom at rest. Specifically, describe the difference in photon energy required for an atom moving towards the light source versus an atom moving away from the light source.

4.5 A vapor cell has strontium-84 atoms. A strontium-84 atom has 38 protons and 46 neutrons (notice $38 + 46 = 84$). The mass of a strontium-84 atom is 1.393×10^{-25} kg.

(a) If the temperature of the vapor cell is 350 K, what is the full width at half maximum of the Maxwell-Boltzmann velocity distribution?

(b) There is a transition from the ground state to an excited state at 650.5032 THz. There are no other energy levels nearby, so you can treat this transition as a two-level atom. What is the Doppler width for this spectral feature? Give your answer in MHz.

(c) Sketch the transmission plot of a laser beam as it passes through a vapor cell held at 350 K. You can pick any amplitude you want for the Doppler feature. Hint: You should be using your answer from part b) in this sketch.

4.6 Starting with Eq. 4.14, derive Eq. 4.15.

4.7 (The Full Doppler Shift Formula: Moving Observers and Sources) In non-relativistic physics, the formula for the Doppler shift for a light wave is:

$$f_{obs} = \left(\frac{c \pm v_{obs}}{c \mp v_{em}} \right) f_{em}, \tag{4.16}$$

where f_{obs} is the frequency measured by the observer and f_{em} is the frequency emitted by the source. v_{obs} is the speed of the observer relative to some background and is always a positive number (it is a speed). It is added to c in the numerator if the observer is moving towards the source and subtracted if the observer is moving away from the source. v_{em} is the speed of the source with respect to that same background and is also always a positive number (it is a speed). It is added to c in the denominator if the source is moving away from the observer and subtracted if the source is moving towards the observer.

(a) Come up with 4 different scenarios for the 4 different sign combinations. For example, what is a scenario where the observer is moving away from the source and the source is moving towards the observer? In this scenario, you would use the formula:

$$f_{obs} = \left(\frac{c - v_{obs}}{c - v_{em}} \right) f_{em}, \tag{4.17}$$

(b) Check to make sure this formula agrees with Eq. 4.2.

(c) Check to make sure this formula agrees with Eq. 4.14.

4.8 Figure 4.8 is a picture of the spectrum from a distant galaxy that you can download from the Sky Server database.[5] The Sky Server ID for this galaxy is

---

[5] Image and data is from the Sloan Digital Sky Survey. Funding for the Sloan Digital Sky Survey (SDSS) has been provided by the Alfred P. Sloan Foundation, the Participating Institutions, the National Aeronautics and Space Administration, the National Science Foundation, the U.S.

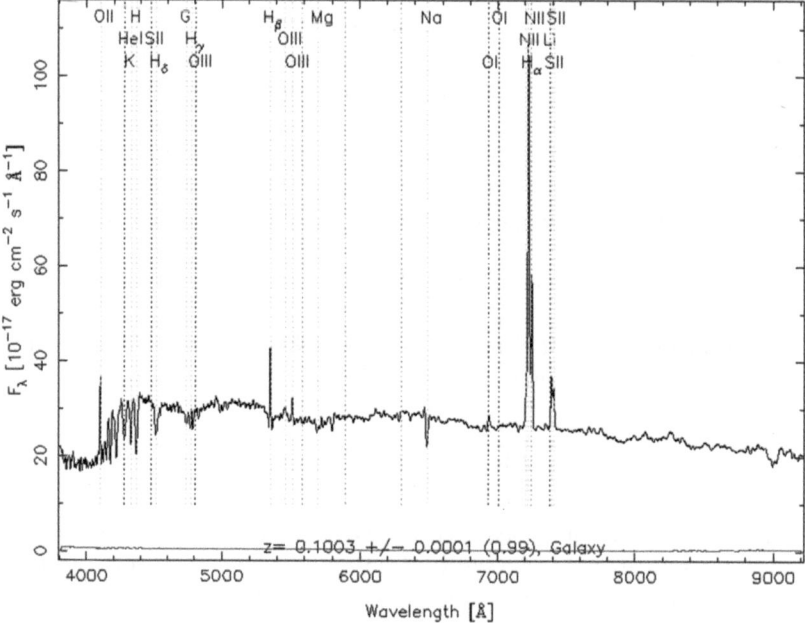

Fig. 4.8 Light collected on earth from galaxy 582102012537667624

Table 4.2 Rest wavelengths
of Hydrogen–Balmer series

Name	Wavelength (angstroms)
H_α (H-alpha)	6562.8
H_β (H-beta)	4861.3
H_γ (H-gamma)	4340.5
H_δ (H-delta)	4101.7

582102012537667624. The galaxy is emitting a number of photons from different elements including hydrogen, oxygen, and magnesium. We are going to focus on the hydrogen lines. On earth, we measure those hydrogen lines to have wavelengths that are given in Table 4.2.

Department of Energy, the Japanese Monbukagakusho, and the Max Planck Society. The SDSS Web site is http://www.sdss.org/.

The SDSS is managed by the Astrophysical Research Consortium (ARC) for the Participating Institutions. The Participating Institutions are The University of Chicago, Fermilab, the Institute for Advanced Study, the Japan Participation Group, The Johns Hopkins University, Los Alamos National Laboratory, the Max-Planck-Institute for Astronomy (MPIA), the Max-Planck-Institute for Astrophysics (MPA), New Mexico State University, University of Pittsburgh, Princeton University, the United States Naval Observatory, and the University of Washington.

Use the data graphed in Fig. 4.8 to estimate the wavelengths of those lines that astronomers measured here on earth and find the speed of galaxy 582102012537667624 relative to the earth.

Saturated Absorption Spectroscopy

5

Abstract

In this chapter, we explore the clever spectroscopy technique known as saturated absorption spectroscopy. This technique is used to remove Doppler profiles from spectroscopic signals. We will learn how saturated absorption spectroscopy works, including the roles of probe and pump beams, and the resulting spectral features. Additionally, we will examine the artifacts, specifically crossover features (V, Λ, and X crossovers), that may appear due to this technique and understand the conditions under which they occur. Practical examples using various atoms, advanced techniques for achieving crossover-free spectroscopy, and potential issues are also discussed.

Learning Goals

By the end of this chapter, you should be able to understand:

- saturated absorption spectroscopy: How we use two counterpropagating lasers to produce a spectrum that looks like the atoms are at 0 Kelvin.
- crossover features: Artifacts of saturated absorption spectroscopy.
- the conditions under which different types of crossover features (V, Λ, X) appear and how to identify them in a spectrum.

5.1   Saturated Absorption Spectroscopy

Saturated absorption spectroscopy is a really neat spectroscopy trick used on a vapor cell with hot atoms that creates a transmission plot with only spectral features from the atoms that are moving perpendicular to the laser ($v_\parallel = 0$). Reread that

W. Raven, *Atomic Physics for Everyone*,
https://doi.org/10.1007/978-3-031-69507-0_5

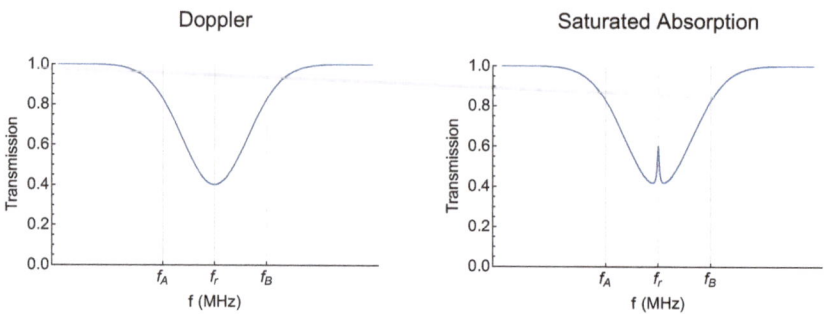

Fig. 5.1 An illustrative example showing the transmission plots for a two-level atom. On the left is the Doppler profile we learned about in Chap. 4. On the right is the transmission plot for a saturated absorption setup

sentence! It is really quite amazing. Suppose the atoms are at 400 K. We know that if we use a single laser beam, we would expect to see a Doppler broadened spectrum from these atoms that is Gaussian in shape. Saturated absorption spectroscopy uses two laser beams, resulting in a small Lorentzian feature on top of the Gaussian shape, as shown in Fig. 5.1. The small Lorentzian feature comes only from those atoms that have zero speed in the direction of the laser.[1]

This is how we do it: we send two laser beams into a vapor cell from opposite directions, see Fig. 5.2. The laser beam that starts on the left and moves to the right has a small amount of power. We call this laser the **probe beam**. In saturated absorption spectroscopy, we measure the transmission of the probe beam. The other laser starts on the right and is moving to the left and has a large amount of power. We will call this laser the **pump beam**. In most experimental setups, the probe beam and the pump beam originate from the same laser. The laser can be split into two paths using, for example, a $\lambda/2$ plate and a polarizing beam splitter, see Sect. 1.5. One would adjust the orientation of the $\lambda/2$ plate so that the probe beam has less power than the pump beam.

Fig. 5.2 Saturated absorption spectroscopy needs two laser beams: a probe beam and a pump beam. Not pictured is the photodiode with which we monitor the transmission of the probe beam

---

[1] Remember that an atom can be moving perpendicular to the laser beam, and it will experience no Doppler effect. Only the velocity component parallel (towards or away) with the laser will contribute to a Doppler shift.

Fig. 5.3 An example of a
Doppler profile simulated
using parameters for a
transition in europium-156
atoms with a temperature of
400 Kelvin

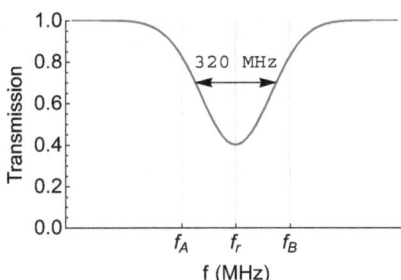

To explore how this technique works, we will use our simple two-level atom. If
the pump beam were not present, we know the transmission of the probe beam looks
like Fig. 5.3. This plot is calculated using the mass of a europium-156 atom and a
vapor cell at 400 Kelvin. The natural linewidth of the transition is about 25 MHz,
which is much smaller than the Doppler width of 320 MHz.

Now, let's add in the pump beam. Our ultimate goal is to determine how the
transmission plot of the probe beam changes with the addition of the pump beam.
Let's start by thinking about the transmission of a laser through the vapor cell when
the laser frequency is at f_A, see Fig. 5.1. Since both the pump and the probe beam
come from the same laser, they have the same frequency. The only difference is that
they are moving in opposite directions. We want to ask the question: Which atoms
interact with each laser beam?

As a reminder, the Doppler shift for an atom moving with velocity component $v_\parallel$
is given by the formula:

$$\Delta f_D = -\frac{v_\parallel}{\lambda} = -\frac{v_\parallel}{c} f_L \rightarrow |v_\parallel| = \Delta f_D \frac{c}{f_L} \tag{5.1}$$

where $v_\parallel$ is negative if the atom is moving towards the laser source and positive if it
is moving away.

I find it useful to use numbers, so let's say that the frequency of the laser is set
to f_A, and f_A is 200 MHz below $f_r = 652.0000$ THz. Since we have two laser
beams moving in different directions, we are going to use the magnitude of $v_\parallel$ and
the descriptors "to the left" and "to the right" for the following discussion.

Let's start with the probe beam, which is moving to the right. Since the frequency
of the laser is below resonance, we know that atoms which interact with the probe
beam have to be moving to the left with a specific $v_\parallel$ so that, according to those
atoms, the Doppler effect shifts the laser frequency into resonance. That means the
atoms would have to be moving at:

$$v_\parallel = (200 \times 10^6 \text{ Hz}) \frac{3 \times 10^8 \text{ m/s}}{652 \times 10^{12} \text{ Hz}} = 92 \frac{\text{m}}{\text{s}} \text{ to the left} \tag{5.2}$$

to absorb light from the probe beam.

Now let's think about the pump beam, which is moving to the left. That means the atoms that absorb light from the pump beam must be moving to the right. The math is the same, but the direction of movement is opposite:

$$v_{\parallel} = (200 \times 10^6 \, \text{Hz}) \frac{3 \times 10^8 \, \text{m/s}}{652 \times 10^{12} \, \text{Hz}} = 92 \, \frac{\text{m}}{\text{s}} \text{ to the right.} \tag{5.3}$$

Spend a few minutes on the above argument to make sure it all makes sense.

Here is the important take home message: When the frequency of the laser is at f_A, **different** atoms interact with the probe beam and the pump beam. Both lasers are losing photons, but they are losing photons to **different** atoms. Since we are monitoring the probe beam transmission, the probe beam transmission is the same whether the pump beam is on or off. Again, this is a very important concept so make sure it makes sense before moving on.

Your turn! The laser frequency is now at frequency f_B, which we will assume is 200 MHz higher than $f_r = 652.0000 \, \text{THz}$. What velocity does an atom need to have to absorb light from the probe beam? From the pump beam? The answers are in the footnotes.[2]

The conclusion for when $f_L = f_B$ is the same as when $f_L = f_A$: When the frequency of the laser is at f_B, the probe beam transmission is the same whether the pump beam is on or off. The trick happens when the laser frequency is at f_r. The Doppler shift is 0, so both the probe and the pump beams interact with the **same** atoms. When the laser frequency is at f_A or f_B (or any frequency except f_r), the pump and the probe lasers interact with **different** atoms. When the laser frequency is at f_r, the two laser beams compete for the **same** atoms.

To explore this more, let's do a thought experiment. First, we either block or turn off the pump beam so that there is only a probe beam. The probe beam frequency is set to the resonance frequency, and we'll assume it hits a single atom at rest. Let's say the probe beam has 10 photons that pass by the atom for every lifetime of the excited state. From those 10 photons, the atom absorbs 1 photon reducing the probe beam transmission to 9 photons; this is a 10% reduction in probe beam transmission. Now we turn the pump beam back on. The pump beam has more power than the probe beam. Let's say the pump beam provides an additional 990 photons. The atom will randomly pick 1 photon from a possible 1000 photons (10 from the probe and 990 from the pump). Most likely the atom is going pick a photon from the pump beam. Since all 10 photons make it through, the transmission of the probe beam is larger when the pump beam is on. Every once in a while, the atom will randomly absorb from the probe beam, decreasing its transmission percentage. However, the transmission of the probe beam is, on average, larger when the pump beam is present. If we increase the number of atoms in the vapor cell, each atom with $v_{\parallel} = 0$ will randomly absorb from either the pump beam or the probe beam.

---

[2] Probe: 92 m/s (to the right); Pump: 92 m/s (to the left); notice the directions are switched from when the laser frequency was f_A.

The conclusion is: *When the laser frequency matches the resonance frequency, the transmission of the probe beam is larger when the pump beam is present.*

What is really important is that this only happens for the atoms that have $v_\parallel = 0$. For any other velocity, the probe beam transmission is exactly the same whether the pump beam is on or off. Let's recap all of this in a table. To make things easier, we are going to define "moving to the left" (towards the probe beam) as negative and "moving to the right" as positive.

f_L	Velocity of atoms needed to absorb from probe beam.	Velocity of atoms needed to absorb from pump beam.	How does the pump beam change the transmission of the probe?
f_A	-92 m/s	$+92$ m/s	It doesn't
f_r	0 m/s	0 m/s	Transmission increases
f_B	$+92$ m/s	-92 m/s	It doesn't

Summary
If the laser frequency is not on resonance, the probe beam and pump beam are interacting with different atoms. In other words, the probe beam is losing photons to different atoms than the pump beam. On resonance, the two lasers compete for the same atoms, which results in less photons being absorbed from the probe beam.

A simulation of the transmission of the probe beam with the pump beam off (left plot) and with the pump beam on (middle) is shown in Fig. 5.4. If we subtract the two plots (right), we are left with a spectral feature with a full width half maximum equal to the natural linewidth of the transition.[3] This is the same plot as the absorption plot from the thought experiment that we did in Sect. 3.1, which was an absorption plot

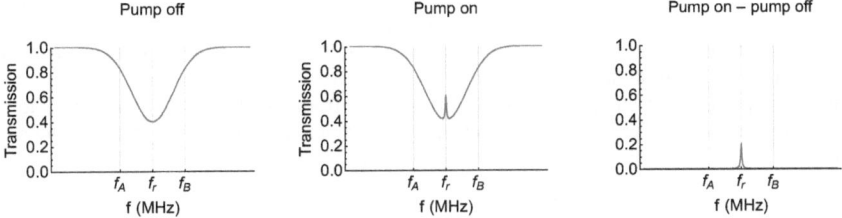

Fig. 5.4 The transmission plots of the probe beam for a 2 level atom with just a probe beam (left), the probe beam and a pump beam (middle), and the difference between the two transmission plots (right)

---

[3] Assuming that the width is not broadened from some other effect like power broadening.

for atoms at 0 Kelvin. We call this plot a **saturated absorption plot**. Neat trick, huh!

5.2   Crossovers

Saturated absorption spectroscopy is super cool.[4] It allows us to use hot gas and still produce spectral features as if all of the atoms were frozen in place (absolute zero or 0 K). However, there is a trade off if there are multiple ground or excited states, and that trade-off is additional fake spectral features in our spectrum called crossovers. There are three types of crossovers: V crossovers, Λ crossovers (the Greek letter capital lambda, so we call them "Lambda crossovers"), and X crossovers. The reason for the names should be clear as you examine the energy level diagrams for each in Fig. 5.5. I want to point out that the labels, $F = 4$, $F = 5$, etc. have meaning that we explore in Chaps. 7 and 8. Even though we haven't connected those labels to physics yet, I wanted to remind you about "The Rule" from Eq. 3.15:

The Rule

$$\Delta F = -1, \ 0, \ +1$$
$$F = 0 \nrightarrow F = 0 \tag{5.4}$$

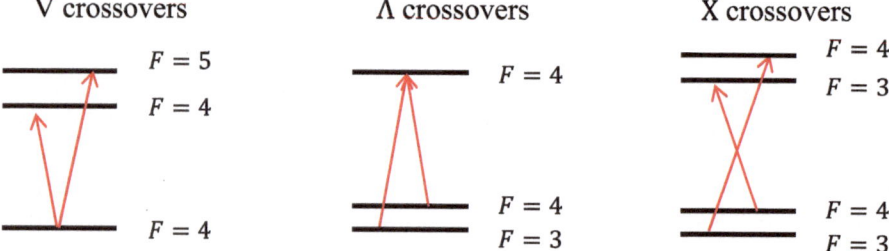

Fig. 5.5 The three types of crossovers. A V crossover is due to the probe beam and pump beam exciting atoms from the same ground state to two different excited states. For a V crossover, the pump beam "steals" atoms from the probe beam. A Λ crossover is due to the probe beam and pump beam exciting atoms from the two different ground states to the same excited state. A X crossover is due to the probe beam and pump beam exciting atoms from two different ground states to two different excited states. For both Λ and X crossovers, the pump beam "gives" atoms to the probe beam

---

[4] Yay puns!

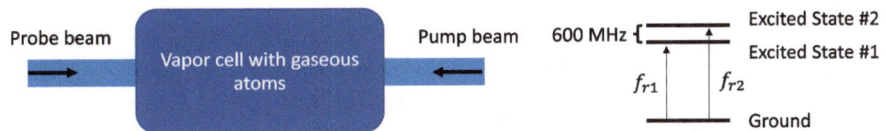

Fig. 5.6 The experimental setup and energy levels to think about V crossovers

5.2.1 *V* Crossovers

Suppose you have an atom with one ground state that can be excited to two different excited states, as shown in Fig. 5.6. The two excited states have resonance frequencies f_{r1} and f_{r2}. Using the arguments explored in Sect. 5.1, you might expect transmission plots shown in Fig. 5.7. In the "Pump off" plot, I also plotted the individual Doppler profiles for both transitions (red and blue dashes). If you add these two together you will get the black curve. With the pump beam on, you might (correctly) expect to get Lorentzian features at laser frequencies f_{r1} and f_{r2}.

That prediction is close, but not quite correct. If you do the experiment, you will find that you have an additional Lorentzian shaped spectral feature exactly halfway between f_{r1} and f_{r2}. This extra feature is called a *V* crossover. Let's explore why this happens with an example. I always find it easier to use numbers, so let's say that f_{r2}-$f_{r1} = 600$ MHz. The feature occurs when the laser frequency is set to 300 MHz above f_{r1} and 300 MHz below f_{r2}. We are also going to define an atom moving right as positive velocity and an atom moving left as negative velocity.

Your turn: Having the laser frequency precisely between f_{r1} and f_{r2}, calculate the velocity that an atom would need in order to absorb from the pump beam to excited state #1, from the probe beam to excited state #1, from the pump beam to excited state #2, and from the probe beam to excited state #2. Use $\lambda = 500$ nm for the math. Make sure you have an answer before moving on. Here is a crossword puzzle to separate the question and answer.

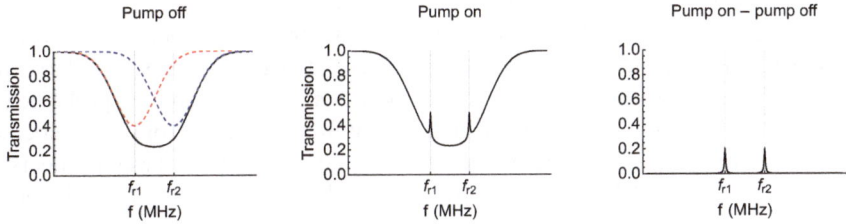

Fig. 5.7 A very reasonable, but incorrect guess for a saturated absorption spectrum for an atom with one ground state and two excited states

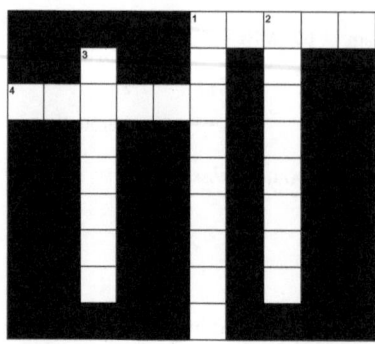

Across:

1   An experimental device that provides photons
4   A particle of light

Down:

1   Inversely proportional to the lifetime of a state
2   ____ feature
3   $\frac{v_1}{c} f_L$ is the formula for the ____ shift

	Velocity of atoms needed to absorb from probe beam.	Velocity of atoms needed to absorb from pump beam.
Excited state #1	+150 m/s	−150 m/s
Excited state #2	−150 m/s	+150 m/s

First, notice that the pump and probe beams are exciting different atoms to excited state #1. The two beams are also exciting different atoms to excited state #2. Specifically, for an atom to be excited by the probe to excited state #1 it would have to be moving away from the probe beam at +150 m/s. An atom would have to be moving away from the pump beam at −150 m/s to be excited by the pump beam to excited state #1. The pump beam and probe beam are interacting with different atoms; nothing new here.

Now notice that the pump beam is trying to excite atoms moving at +150 m/s to excited state #2 while the probe beam is trying to excite those same atoms to excited state #1. Those atoms that are moving at +150 m/s get to pick which laser to absorb from! They can absorb from the probe beam and be excited to excited state #1 or absorb from the pump beam and be excited to excited state #2. The atom is more likely to absorb a photon from the pump beam leaving fewer atoms for the probe to interact with. Even though the two lasers are trying to excite to different states, the pump beam still "steals" atoms from the probe beam meaning the transmission of the probe beam will increase at that frequency. Similarly, the atoms moving at −150 m/s also get to pick between the pump and the probe beam. Therefore, there will be an additional spectra feature that comes from two velocities of atoms (+150 m/s and −150 m/s) when the frequency of the laser is precisely between f_{r1} and f_{r2}. With this new information, the transmission plot of the probe beam has three spectral features: two that correspond to the actual frequencies of the transitions and a third exactly halfway between that we call a crossover peak, see Fig. 5.8.

The two real transitions come from atoms that have no velocity components in the direction of the laser beams. The crossover peak comes from atoms that are moving. I think now is a good time to remind everyone that the amplitudes of the peaks in the above graphs are completely made up. Because there are two sets of

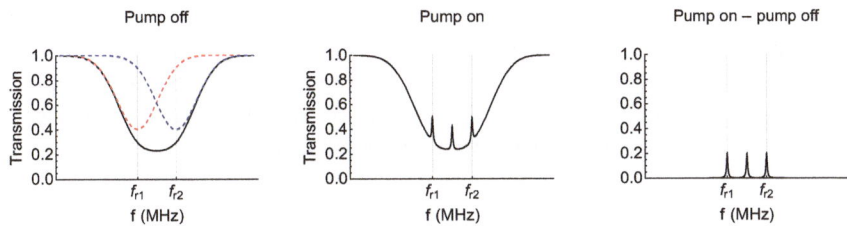

Fig. 5.8 A more accurate simulation of a saturated absorption spectrum with one ground state and two excited states. The amplitudes of the spectral features are made up; in a real experiment, all three features will have different amplitudes

atoms contributing to the crossover peak ($v = +150$ m/s and $v = -150$ m/s), the crossover peak often turns out to be larger than the actual transitions. Also, the amplitude for resonance 1 will not be the same as the amplitude for resonance 2.

▶   **Important Comment**   If the two transitions are separated such that the Doppler profiles of each transition are separated, you will not have any crossovers because there are no atoms moving with the correct speeds to cause the crossover feature, see Fig. 5.9.

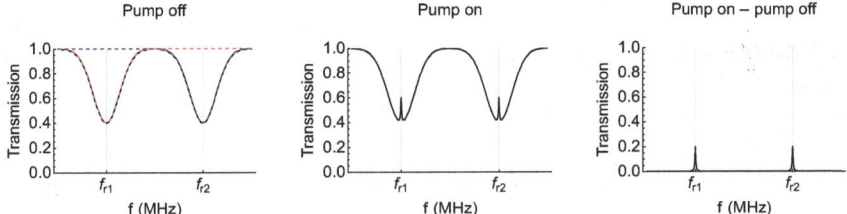

Fig. 5.9 A simulation of a saturated absorption spectrum with one ground state and two excited states, but the two excited states are separated by a large energy. There are no atoms with the correct velocity to create the V crossover. The amplitudes of the spectral features are made up; in a real experiment, the features will have different amplitudes

If the vapor cell was heated to increase the Doppler width, the crossover peak would return, see Fig. 5.10.

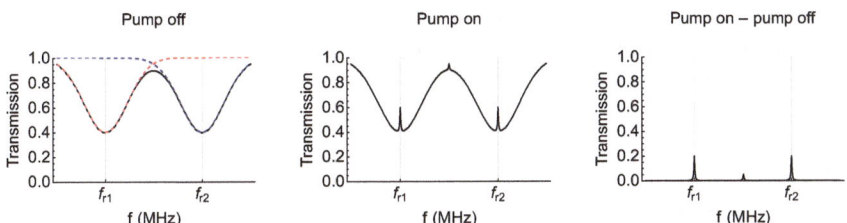

Fig. 5.10 Now the vapor cell is heated so there are a few atoms with the correct velocity needed to create a V crossover

Summary
If there are (1) two excited states and one ground state and (2) the Doppler profiles for the two individual transitions are overlapping one another, there will be a *V* crossover directly between the two transitions.

The speed of atoms needed to create a crossover feature can be derived from the formula for the Doppler shift. The speed needed is:

$$|v_{\parallel}| = \lambda \frac{|f_{r1} - f_{r2}|}{2}. \tag{5.5}$$

If there aren't any atoms in the vapor cell with that speed, there won't be a crossover feature. As with many things in experimental science, there are trade-offs to saturated absorption spectroscopy. While saturated absorption spectroscopy gives us really narrow spectroscopy features, it gives us more features to deal with. Fortunately, we know precisely where those crossovers will be.

We can add a third excited state to the system.[5] Let's call the resonant frequencies f_{r1}, f_{r2}, and f_{r3}. We will get 3 crossovers features calculated using the same logic as above. One crossover will be directly between f_{r1} and f_{r2} (i.e., $(f_{r1}+f_{r2})/2$), one between f_{r1} and f_{r3} (i.e., $(f_{r1}+f_{r3})/2$), and one between f_{r2} and f_{r3} (i.e., $(f_{r2}+f_{r3})/2$). In total, saturated absorption spectroscopy on an atom with one ground state and three excited states will have 6 spectral features: 3 real features and 3 crossovers.

5.2.2   Λ Crossovers and *X* Crossovers

Λ crossovers and *X* crossovers, see Fig. 5.5, both occur when the pump beam excites an atom with a particular velocity, such that, upon decay to a different ground state, that atom has the correct velocity to be excited by the probe beam. As a reminder, *V* crossovers occur because an atom gets to pick between absorbing a photon from the pump beam and the probe beam. When an atom picks the pump beam over the probe beam, the transmission of the probe beam increases resulting in a bump on the transmission or absorption plot. Λ crossovers and *X* crossovers both occur because the pump beam puts more atoms in the probe beam's path. As such, the transmission of the probe beam decreases resulting in a dip on the transmission or absorption plot. Like *V* crossovers, this feature occurs when the laser frequency is precisely between the two resonance frequencies:

---

[5] We can't add more than that for a single ground state due to "The Rule".

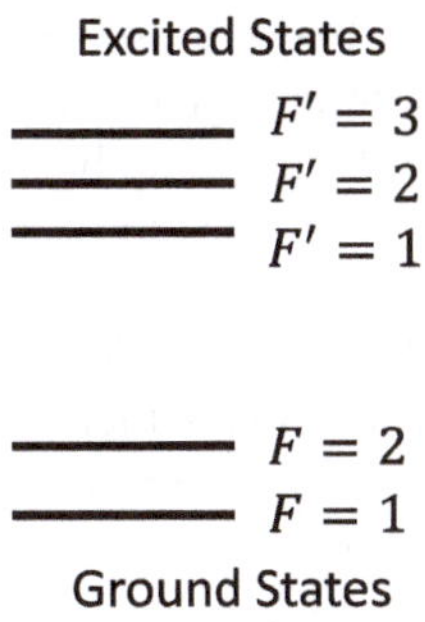

Fig. 5.11 An atom with two ground states and three excited states

Excited States

$\underline{\qquad}$ $F' = 3$
$\underline{\qquad}$ $F' = 2$
$\underline{\qquad}$ $F' = 1$

$\underline{\qquad}$ $F = 2$
$\underline{\qquad}$ $F = 1$

Ground States

$$f_{\text{cross}} = \frac{f_{r1} + f_{r2}}{2} \tag{5.6}$$

As before, the vapor cell needs atoms with a speed

$$|v_{\|}| = \lambda \frac{|f_{r1} - f_{r2}|}{2}. \tag{5.7}$$

to create these features.

There is one important difference for X crossovers. For V crossovers and Λ crossovers, the pump and the probe beam are interchangeable. Consider an atom that has two ground states and three exited states, see Fig. 5.11. We pick the ground states to have labels $F = 1$ and $F = 2$ and the excited states to have labels $F' = 1$, $F' = 2$, and $F' = 3$.[6] As a reminder, "The Rule" is that an atom can be excited as long as $\Delta F = 1$, 0, or -1 with the exception $F = 0 \nrightarrow F = 0$. Suppose we have a Λ crossover that comes from the two transitions $F = 1 \rightarrow F' = 2$ and $F = 2 \rightarrow F' = 2$. To add numbers, let's say the vapor cell needs atoms with speed $v_{\|} = +35$ m/s or -35 m/s to produce this crossover. It doesn't matter if the pump beam is exciting the first transition or the second. If the pump beam is exciting the first transition, it is "pumping" atoms with $v_{\|} = +35$ m/s from the $F = 1$ ground state to the $F = 2$ ground state via the $F' = 2$ excited state. The probe beam is then exciting those extra atoms on the $F = 2 \rightarrow F' = 2$ transition. If the pump beam is exciting the second transition, it is "pumping" atoms with $v_{\|} = -35$ m/s from the $F = 2$ ground state into the $F = 1$ ground state via the $F' = 2$ excited state. The probe beam is then exciting those extra atoms on the $F = 1 \rightarrow F' = 2$ transition. The important thing to notice here is that the excited state of both transitions can decay into either ground state. We say that there are two "velocity classes" of atoms that are contributing to that crossover feature: $+35$ m/s and -35 m/s.

---

[6] I added primes to the excited states to help us distinguish between the ground states and the excited states.

For X crossovers, there are some situations where the two transitions cannot be interchanged. Consider the two transitions: $F = 1 \rightarrow F' = 2$ and $F = 2 \rightarrow F' = 3$. If the pump beam is exciting the first transition, it is "pumping" atoms with, say, $v_\parallel = +25$ m/s from the $F = 1$ ground state into the $F = 2$ ground state via the $F' = 2$ excited state. The probe beam is then exciting those extra atoms on the $F = 2 \rightarrow F' = 3$ transition. However, if the pump beam is exciting the second transition, which would be the atoms with $v_\parallel = -25$ m/s, $F' = 3$ cannot decay into the $F = 1$ ground state. So, the probe beam transmission, which is exciting atoms on the $F = 1 \rightarrow F' = 2$ transition, is not changed. This crossover only has one "velocity class" that contributes to the crossover, so it tends to be smaller than a crossover with two velocity classes.

We now have the basic building blocks to interpret a spectrum from an atom with as many ground and excited states that we want. If our atom has energy levels as shown in Fig. 5.11, we will have multiple real transitions and multiple crossovers. For the real transitions, an atom in the $F = 1$ ground state can be excited to the $F' = 1$ or $F' = 2$ excited states. An atom in the $F = 2$ ground state can be excited to the $F' = 1$, $F' = 2$, or $F' = 3$ excited states. Each of these five transitions will have a Doppler profile that has a Doppler width associated with it.

A saturated absorption plot will have those five spectral features as well as crossovers. For a crossover to occur, the vapor cell has to have atoms with the velocity needed to create that crossover. To conclude this section, let's recap the three types of crossovers and list the possible crossovers for the atom with the energy states shown in Fig. 5.11:

(1) V crossovers: If there are two excited states that are excited from the same ground state and the Doppler profiles from the individual transitions are overlapping, we will have a crossover whose frequency is directly between the two transitions. Using Fig. 5.11 as an example, V crossovers occur due to interference between:

Transition #1	Transition #2
$F \rightarrow F'$	$F \rightarrow F'$
$1 \rightarrow 1$	$1 \rightarrow 2$
$2 \rightarrow 1$	$2 \rightarrow 2$
$2 \rightarrow 2$	$2 \rightarrow 3$
$2 \rightarrow 1$	$2 \rightarrow 3$

For this example, there are 4 possible V crossovers.

(2) Λ crossovers: If there are two ground states that can be excited to a single excited state and the Doppler profiles from the individual transitions are overlapping, we will have a crossover whose frequency is directly between

the two transitions. Using Fig. 5.11 as an example, Λ crossovers occur due to interference between:

Transition #1	Transition #2
$F \to F'$	$F \to F'$
$1 \to 1$	$2 \to 1$
$1 \to 2$	$2 \to 2$

For this example, there are 2 possible Λ crossovers.

(3) X crossovers: If the pump beam can excite an atom that decays into the ground state for the probe beam and the Doppler profiles from the individual transitions are overlapping, we will have a crossover whose frequency is directly between the two transitions. X crossovers do not share any states. Using Fig. 5.11 as an example, X crossovers occur due to interference between:

Pump	Probe	Notes
$F \to F'$	$F \to F'$	
$1 \to 1$	$2 \to 2$	Interchangeable; two velocity classes
$1 \to 1$	$2 \to 3$	Not interchangeable; one velocity class
$1 \to 2$	$2 \to 1$	Interchangeable; two velocity classes
$1 \to 2$	$2 \to 3$	Not interchangeable; one velocity class

For this example, there are 4 possible X crossovers.

So, in this example, our transmission plot will have up to 15 spectral features. Five of those features will be the real transitions, and the remaining 10 are all crossovers. Again, whether or not those crossovers produce spectral features depend upon there being the correct velocity class of atoms in the sample to produce those features.

5.3   Example with Cesium-133

Cesium-133, which has 55 protons and 78 neutrons, is one of the most studied atoms on the periodic table. Figure 5.12 shows a simplified energy level diagram for a transition that uses 455.6 nm light. The lower state, which has the label 6s $^2S_{1/2}$ (don't worry about what that means right now, we will talk about the physical meaning behind the labeling starting in Chap. 7), has two closely spaced ground states with labels $F = 3$ and $F = 4$ (we will give meaning to these labels in Chaps. 8 and 9). The separation of these two states is just over 9 GHz. In energy units, that would be $hf = (6.626 \times 10^{-34} \text{ Js})(9.192 \times 10^{9} \text{ Hz}) = 6.091 \times 10^{-24} \text{ J} = 38 \, \mu\text{eV}$.

Fig. 5.12 A simplified energy level diagram for the transitions in cesium-133 near 455.6 nm

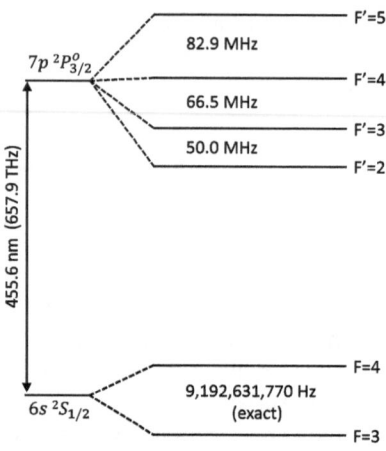

The excited state studied here has four levels. These four levels are far closer together than the two ground state levels. To easily see all of the levels in the figure, the energy spacing scale is different for the ground state and excited state; the energy separation of the two ground states is over 100 times bigger than the excited state separations. Below are 4 questions to work through. Answer the first two questions together before answering the second two questions.

Question #1 What speed does an atom have to have to create a Λ crossover between the two ground states and the $F' = 3$ excited state?

Question #2 Using Eq. 4.5, what temperature would the cesium vapor cell be such that the FWHM of the Maxwell Boltzmann distribution was half of the velocity for Question #1? The mass of cesium-133 is $m = 2.207 \times 10^{-25}$ kg. The answers are below this fun anagram puzzle.

#1: $|v_\|| = \lambda \frac{|f_{r1} - f_{r2}|}{2} = (455.6 \times 10^{-9} \text{ m}) \frac{9,192,631,770 \text{ Hz}}{2} = 2094 \text{ m/s} \rightarrow \frac{|v_\||}{2} = 1047$ m/s.

#2: $v_{\text{FWHM}} = 2.355 \sqrt{\frac{k_B T}{m}} \rightarrow T = \left(\frac{v_{\text{FWHM}}}{2.355}\right)^2 \frac{m}{k_B} = \left(\frac{1047 \text{ m/s}}{2.355}\right)^2 \frac{2.207 \times 10^{-25} \text{ kg}}{1.38 \times 10^{-23} \text{ J/K}} = 3161$ K.

Fun Fact

This energy separation is how we define 1 second! Imagine you had a pendulum that made exactly 9,192,631,770 oscillations in 1 second. Replace that pendulum with a cesium atom and you have the official definition of a second.

Anagram Fun
Rearrange the letters in "cesium" to make a new 6 letter word.
How many 5 letter words can you create from the word "cesium"? (I found one, but an online anagram solver found two!)
How many 4 letter words can you create from the word "cesium"?

This is really hot! For reference, room temperature is about 300 K. In short, we don't have to worry about Λ crossovers (or X crossovers).

Question #3 But what about V crossovers? What velocity does an atom have to have to create a V crossover between the $F = 4$ ground state and the $F' = 4$ and $F' = 5$ excited states?

Question #4 Assuming room temperature, $T = 300$ K, find the FWHM of the Maxwell Boltzmann distribution. What do you conclude? The answers are in the footnotes.[7]

For V crossovers, $|v_\parallel|$ is well within the full width half maximum of the Maxwell-Boltzmann velocity distribution. So, we are definitely going to have V crossovers. However, the $|v_\parallel|$ needed for Λ crossovers and X crossovers is well outside the distribution, so we won't see any Λ crossovers or X crossovers. Figure 5.13 is a saturated absorption plot between the $F = 4$ ground state and the $F' = 3$, $F' = 4$, and $F' = 5$ excited states. The three labeled peaks are the real transitions. Notice there are additional Lorentzian features exactly halfway between

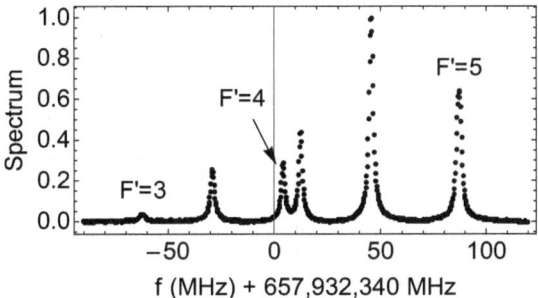

Fig. 5.13 Experimental data taken by my research group showing a saturated absorption plot from the $F = 4$ ground state of cesium-133 to the $F' = 3$, $F' = 4$, and $F' = 5$ excited states, see reference [1]. There are six spectral features. Three of them are real transitions and three are V crossovers

---

[7] #3: $|v_\parallel| = \lambda \frac{|f_{r1} - f_{r2}|}{2} = (455.6 \times 10^{-9} \text{ m}) \frac{8.29 \times 10^6 \text{ Hz}}{2} = 18.9$ m/s.

#4: $v_{\text{FWHM}} = 2.355\sqrt{\frac{k_B T}{m}} = 2.355\sqrt{\frac{(1.38 \times 10^{-23} \text{ J/K})(300 \text{ K})}{2.207 \times 10^{-25} \text{ kg}}} = 322$ m/s. There are definitely atoms in the vapor cell to make this crossover!

any two real transitions. Also notice that all of the amplitudes are different. The crossover between $F' = 4$ and $F' = 5$ is really big while the real transition from the $F = 4$ ground state to the $F' = 3$ excited state turns out to be really small. The peak directly to the right of $F' = 4$ is the V crossover between $F' = 3$ and $F' = 5$. Even if this plot wasn't labeled, we can still figure out which features are the real transitions and which are the crossovers. We just look for the features directly between two other features to find the crossovers. Also, the peaks at the smallest and largest frequency values have to be real transitions; a crossover has to be between two real transitions.

5.4   Oxygen-16: A Spectrum Missing a Crossover

A lot of laser spectroscopy is done from the ground state to an excited state. However, laser spectroscopy can also be performed between two excited states. Figure 5.14a shows a simplified Grotrian diagram for a transition in neutral atomic oxygen-16 (8 protons and 8 neutrons). The lower state, which we give the label $J = 2$, can be excited to three different excited states, which we give the labels $J' = 1$, $J' = 2$, and $J' = 3$. Like in the previous examples, just consider these labels for now. The Rule for these transitions are the same as before, we just replace F with J: $\Delta J = -1, 0,$ or $+1$ with the exception that $J = 0 \not\rightarrow J' = 0$.[8] We will explore what the labels actually mean in Chaps. 7, 8, and 9. Oxygen, in its natural form, is a molecule composed of two oxygen atoms. A discharge (basically think

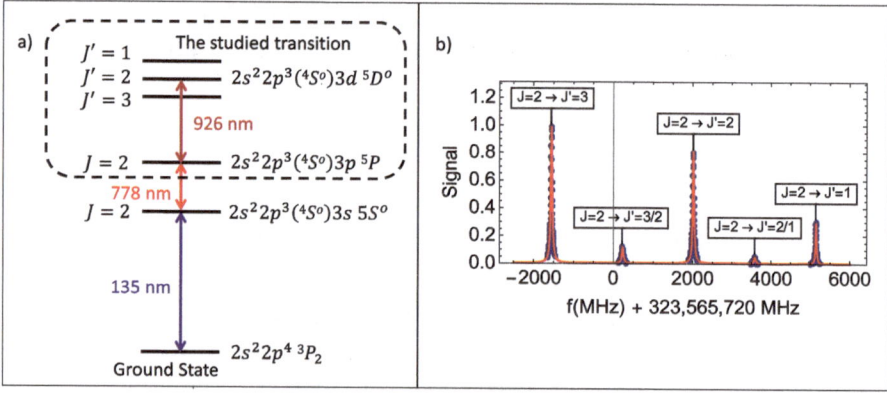

Fig. 5.14 (**a**) A simplified energy level diagram for a spectroscopic study in atomic oxygen-16 near 926 nm. The spectrum is taken between two excited states, which I call the lower state and the upper state. (**b**) Experimental data of a saturated absorption spectroscopy spectrum from the $J = 2$ lower state of oxygen-16 to the $J' = 1$, $J' = 2$, and $J' = 3$ upper states. There are only five spectral features because the vapor cell wasn't hot enough for the V crossover created by the $J = 2 \rightarrow J' = 1$ and $J = 2 \rightarrow J' = 3$ transitions; the Doppler profiles for these two transitions did not overlap

---

[8] A full list of the rules that need to be satisfied for an electron to transition between two atomic states is given in Appendix C.

about a "neon tube" filled with oxygen molecules) can be used to both dissociate the molecule into neutral atomic oxygen as well as excite the electrons into a variety of excited states. Most of the atoms are not in the $J = 2$ lower state, but there are enough for us to do spectroscopy. It should be noted that the $J = 2$ lower state also has a lifetime of about 27 ns, so the discharge needs to continually repopulate the lower state for us to do spectroscopy. Discharges are also typically hotter than room temperature.

Next, take a look at Fig. 5.14b. There is a very visible V crossover created by the large $J = 2 \rightarrow J' = 3$ (real) transition and medium sized $J = 2 \rightarrow J' = 2$ (real) transition. Those transitions are about 3500 MHz apart, but the Doppler profiles of these individual transitions are large enough to create a crossover. This crossover is labeled as $J = 2 \rightarrow J' = 3/2$.[9] The V crossover with the label $J = 2 \rightarrow J' = 2/1$ is created by the medium $J = 2 \rightarrow J' = 2$ and the small $J = 2 \rightarrow J' = 1$ transitions. It is also quite visible, although not as big as the $J = 2 \rightarrow J' = 3/2$ V crossover. Those two transitions are about 3200 MHz apart. We did not see a V crossover created by the $J = 2 \rightarrow J' = 3$ and $J = 2 \rightarrow J' = 1$ transitions, which would have the label $J = 2 \rightarrow J' = 3/1$. Those two transitions are about 6500 MHz apart. Because they are so far apart, the individual Doppler profiles don't overlap resulting in no crossover feature.

5.5 Example with Europium-151

Let's explore a more complex example using europium-151. Consider a transition from the ground state to an excited state that we are going to call the $J' = 5/2$ excited state. Because of nuclear spin, both the ground state and excited state have 6 closely spaced hyperfine levels, see Fig. 5.15. As a reminder, if the nucleus had no angular momentum, there would be a single energy level called the center of gravity that would be located at 0 for both energy level diagrams. The frequency difference between the center of gravity of the excited state and that of the ground state is called the center of gravity frequency. Due to the closely spaced levels, there are a lot of possible transitions and a lot of possible crossovers.

Figure 5.16 is a simulation of a transmission plot with just a probe beam (no pump beam) for a vapor cell with a temperature of 400 K. I plotted the individual transition Doppler broadened spectral features with blue-dashed lines. If you add up the blue curves, you get the black curve. The red vertical lines are the transition frequencies. Because of the temperature of the vapor cell, the individual transitions cannot be resolved. So, this is a good candidate for saturated absorption spectroscopy. The saturated absorption plot will have many transitions (15 of them) and many crossovers (up to 62 of them!). Unfortunately, many of these

---

[9] Just to be clear, that 3/2 is not the fraction equivalent to 1.5. It is meant to convey "a V crossover where the two excited states are $J' = 3$ and $J' = 2$."

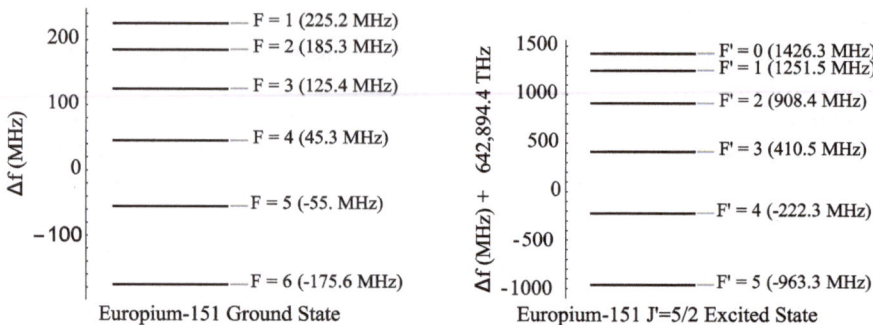

Fig. 5.15 Left: The 6 hyperfine energy levels for the ground state of europium-151. The 6 hyperfine energy levels for a particular excited state that is about 642.9 THz (466.3 nm) above the ground state. The numbers listed for Δf are with respect to the center of gravity

spectral features overlap with each other. Figure 5.17 shows experimental results for a saturated absorption plot collected by my research group on this transition in europium-151. Look how complicated the spectrum is! Although it is a complicated plot, there are spectral features that we can try to attribute to each transition or crossover. Our job, as experimentalists, is to extract as much information as we can from these plots.

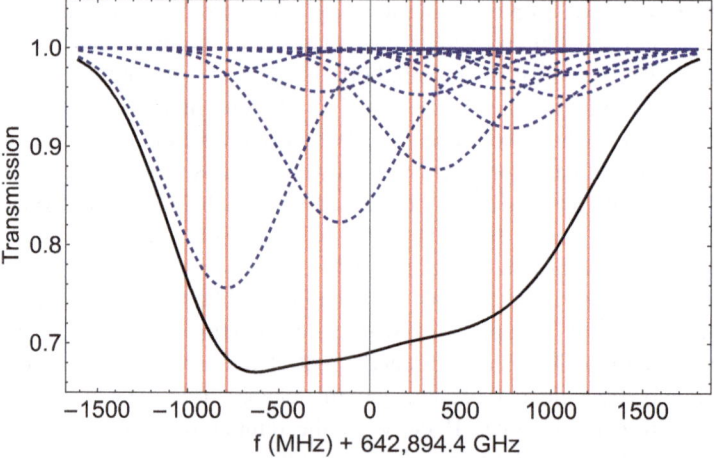

Fig. 5.16 A simulation of a transmission plot with a probe beam (no pump beam) traveling through a vapor cell of europium-151 atoms held at 400 K. There are 15 transitions in total. The Doppler profile for each transition is shown in blue-dashed lines. Some of the amplitudes are quite small and not really visible by eye in this plot. The sum of all the individual Doppler profiles is the black curve, which is what we would measure in the lab. The red vertical lines indicate the center of each transition. As you can see, no single spectral feature can be resolved

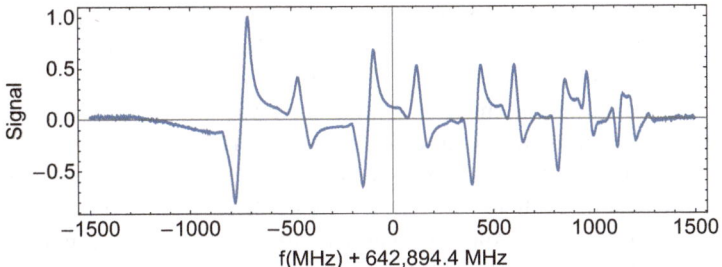

Fig. 5.17 Experimental results of performing saturated absorption spectroscopy on europium-151 atoms, see reference [2]. There are 15 real spectral features and up to 62 crossover features for a total of 77 possible spectral features!

5.6   Extra: Crossover-Free Spectroscopy

Crossovers can be problematic because they introduce additional features into the spectrum. Many times, those crossover features overlap each other or overlap the features from real transitions. So, it isn't too surprising that spectroscopists developed methods of getting sub-Doppler features without crossovers. The simplest idea is to use an atomic beam, see Fig. 5.18.

An atomic beam is created by taking a sample, placing it in a vacuum-compatible oven,[10] and heating the oven. The oven has a small hole to allow the atoms to escape. After the oven, metal pieces called collimators are typically used to block any atoms diverging at large angles. The ideal spectroscopy experiment would have an atomic

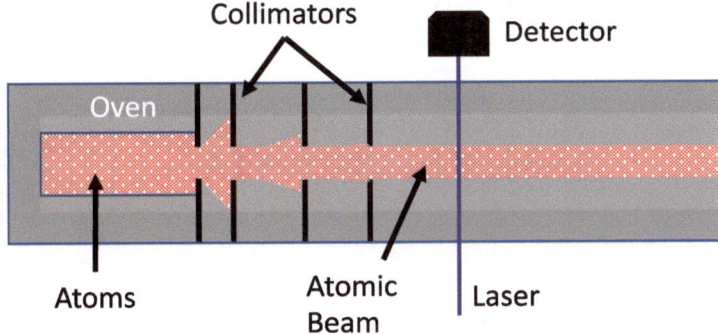

Fig. 5.18 A sketch of how experimentalists use an oven and collimators to make a collimated atomic beam. Since the atoms are not moving vertically, if we sent a laser perpendicular to the atomic beam, then $v_\parallel = 0$

---

[10] Assuming the atoms are a solid at room temperature. For gaseous molecules, the oven is replaced with a discharge to dissociate the molecules into atoms.

beam with zero divergence, resulting in a column of atoms exiting the oven. In practice, there will always be some divergence of the atomic beam.

The laser beam intersects perpendicular to the atomic beam. In this experimental design, there are no atoms moving towards or away from the laser so there are no Doppler shifts and there are no crossovers. This type of setup does have a few drawbacks. The first is that you really need to make sure the laser is perpendicular to the atomic beam. If there is a small angle, there will be no atoms moving perpendicular to the laser. And, you can't really tell if there is a non-zero angle either. You still have atoms absorbing from the laser, but they will all be absorbing at the Doppler shifted frequency. So, the spectrum looks the same, but the resonant frequency is off. A common technique to address this issue is to perform saturated absorption spectroscopy on the atomic beam. The other issue you have to deal with is that the atomic beam is never perfectly collimated. Often times, the atomic beam will be diverging more in one direction than the other. That will cause an asymmetry in the spectral signal, even when using saturated absorption spectroscopy.

Another clever method for doing spectroscopy is to have two laser beams that are traveling in the same direction, see Fig. 5.19. Unlike typical saturated absorption spectroscopy, the two laser beams have independent frequency control. In saturated absorption spectroscopy, the pump and probe beams come from the same laser, so changing the frequency of the laser changes the frequency of both the pump and the probe beams. In this setup, the frequency of laser #1 is going to be fixed to a transition, and the frequency of laser #2 is scanned. The transmission of laser #1 is what we monitor. Laser #2 also has more laser power (a higher saturation parameter).

One obstacle for this experimental setup is that you need two lasers, which can be expensive. The other is that laser #1 has to be at the resonance frequency for one of the transitions. If the frequency of laser #1 does not perfectly match a resonant frequency, the frequency scale of your spectrum will be off.

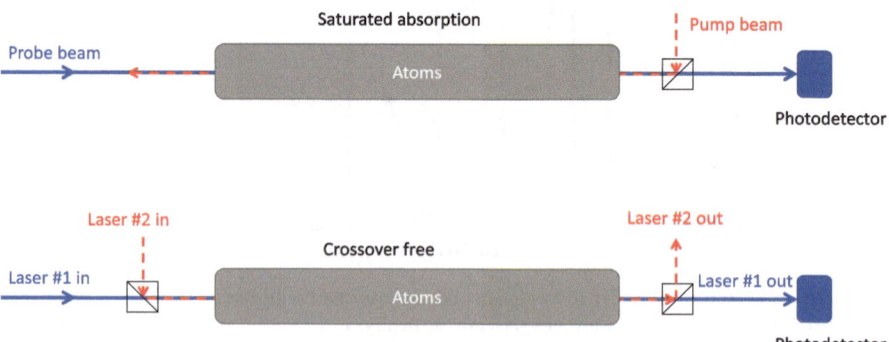

Fig. 5.19 A comparison between a saturated absorption spectroscopy experimental setup and a crossover-free setup. The crossover-free setup requires two separate lasers

Fig. 5.20 A Grotrian diagram to explore crossover-free spectroscopy. Laser #1 has a frequency that exactly matches the $F = 5 \rightarrow F' = 5$ transition frequency. The frequency of Laser #2 is smoothly scanned from a frequency that is too small to excite any resonant transition to too large

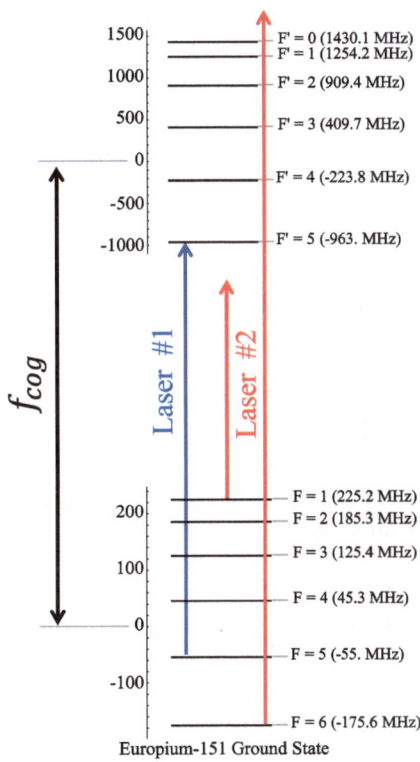

$$F' = 0 \ (1430.1 \text{ MHz})$$
$$F' = 1 \ (1254.2 \text{ MHz})$$
$$F' = 2 \ (909.4 \text{ MHz})$$
$$F' = 3 \ (409.7 \text{ MHz})$$
$$F' = 4 \ (-223.8 \text{ MHz})$$
$$F' = 5 \ (-963. \text{ MHz})$$

f_{cog} Laser #1   Laser #2

$$F = 1 \ (225.2 \text{ MHz})$$
$$F = 2 \ (185.3 \text{ MHz})$$
$$F = 3 \ (125.4 \text{ MHz})$$
$$F = 4 \ (45.3 \text{ MHz})$$
$$F = 5 \ (-55. \text{ MHz})$$
$$F = 6 \ (-175.6 \text{ MHz})$$

Europium-151 Ground State

To better understand how the two-laser spectroscopy set up works, consider the following problem on producing a crossover-free spectrum that looks like a spectrum at 0 Kelvin. Figure 5.20 shows a Grotrian diagram for a transition in europium-151. Laser #1 has a frequency that is fixed to the $F = 5 \rightarrow F' = 5$ transition. Laser #2 is going to scan from a frequency below the $F = 1 \rightarrow F' = 5$ transition to above the $F = 6 \rightarrow F' = 0$ transition. Note that neither of these transitions are allowed. I'm just giving an f_{min} and an f_{max} for our frequency scan. For these transitions, the wavelength of light is around $\lambda = 466$ nm. We will monitor the transmission of Laser #1 as a function of frequency for Laser #2.

(a) Considering only the atoms moving with $v_{\parallel} = 0$, explain why scanning the frequency of Laser #2 across the $F = 5 \rightarrow F' = 5$ transition results in a 0 Kelvin spectral feature on the transmission plot for Laser #1. Will you also get a spectral feature when Laser #2 scans across the $F = 5 \rightarrow F' = 4$ transition?

(b) When Laser #2 scans through the $F = 6 \rightarrow F' = 5$ transition, Laser #2 will excite atoms with $v_{\parallel} = 0$ from the $F = 6$ ground state to the $F' = 5$ excited state. Even though Laser #1 is not resonant with that transition, there will be a spectral feature at that frequency on the transmission plot. Why?

(c) Now consider atoms that are moving at a speed $v_\parallel \approx 344$ m/s towards Laser #1 (and also towards Laser #2). The atoms are moving with the perfect speed to be excited by Laser #1 on the $F = 5 \rightarrow F' = 4$ transition. Laser #2 now has its frequency scanned. How do these atoms affect the transmission plot for Laser #1?

(d) Next, consider atoms that are moving at a speed $v_\parallel \approx 56$ m/s towards Laser #1 (and also towards Laser #2). These atoms are moving with the perfect speed to be excited by Laser #1 on the $F = 6 \rightarrow F' = 5$ transition. How do these atoms affect the transmission plot as the frequency of Laser #2 is scanned?

(e) After considering parts (a) through (d), how many features will our transmission plot have?

Problems

5.1 For each of the following equations, write a brief description of what each equation means.

(a) Equation 5.6
(b) Equation 5.7

5.2 To create a particular crossover, a vapor cell needs atoms that have a speed given by Eq. 5.7. Derive this formula using the Doppler shift equation.

5.3 Consider a transition in an atom with three hyperfine ground states and two hyperfine excited states. The ground states have labels $F = 2$, $F = 3$ and $F = 4$ and the excited states have labels $F' = 3$ and $F' = 4$.

(a) List all of the possible transitions.
(b) List all possible V crossovers.
(c) List all possible Λ crossovers.
(d) List all possible X crossovers.
(e) Optional: Write a computer program to calculate all possible transitions and crossovers for the europium-151 transition studied in Sect. 5.5.

5.4 Figure 5.21 shows energy levels for a transition in rubidium-87. The center of gravity for the ground and excited states are shown on the far left. The ground state labeled $F = 1$ has an energy of -4271.676 MHz with respect to the center of gravity and the ground state labeled $F = 2$ has an energy of $+2563.005$ MHz with respect to the center of gravity. The two ground states are separated by 6834.682 MHz, which is much larger than the width of the Doppler profile for these transitions, which is about 510 MHz, at 300 K. This means there will be no Λ or X crossovers. However, all of the excited states are separated by frequencies smaller than the Doppler width, which means the saturated absorption spectrum will have V crossovers. The natural linewidth for this transition is 6 MHz.

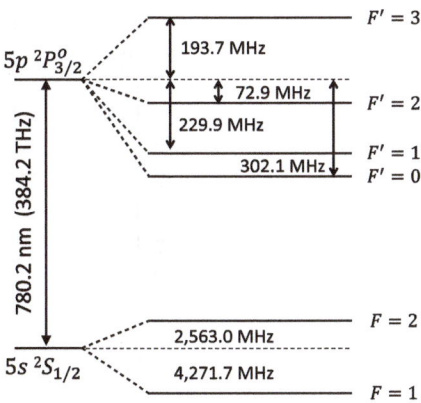

Fig. 5.21 A Grotrian diagram for a transition in rubidium-87

(a) Make a saturated absorption plot using the above energy levels assuming atoms are only in the $F = 1$ ground state. Remember to use The Rule: $\Delta F = -1, 0$, or $+1$ with the exception $F = 0 \nrightarrow F' = 0$. As always, don't worry about the amplitudes of the spectral features. The horizontal axis should be with respect to center of gravity of the excited state. On your plot, label which features are real transitions and which are crossovers.

(b) Make a saturated absorption plot assuming atoms are only in the $F = 2$ ground state.

(c) The plots in part (a) and part (b) are separated by about 6830 MHz, see Fig. 5.22. The 0 on the horizontal axis in Fig. 5.22 is with respect to center of gravity frequency. Let's assume the rubidium atoms are really hot. So hot that the Doppler profiles from the two ground states are overlapping, which means we will have more crossovers. Where on the above graph would the crossover be due to the two transitions $F = 1 \rightarrow F' = 1$ and $F = 2 \rightarrow F' = 1$?

(d) In the scenario outlined in part (c), why would there be no crossovers due to the two ground states and the $F = 0$ excited state?

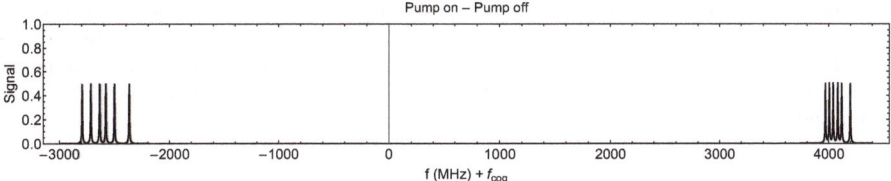

Fig. 5.22 A simulation of a saturated absorption plot (i.e., a pump on - pump off plot) scanning across all possible transitions. The 0 on the horizontal axis is the center of gravity frequency 384.2 THz. The amplitudes for the spectral features are all set to be the same. In a real experiment, the amplitudes will all be different

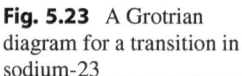

Fig. 5.23 A Grotrian
diagram for a transition in
sodium-23

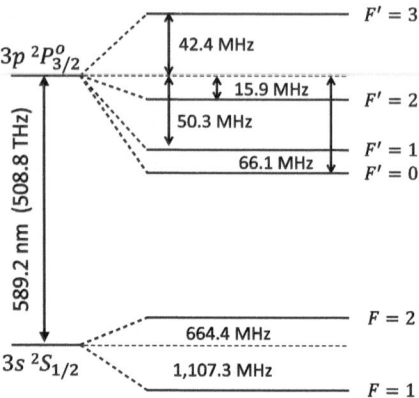

5.5 A transition in sodium-23 has a Grotrian diagram that is very similar to the transition studied in Problem 5.4 for rubidium-87. The Grotrian diagram for the sodium transitions studied in this problem are shown in Fig. 5.23: The difference is that the energy levels are much closer together. The natural linewidth for this transition is 10 MHz.

(a) Make a saturated absorption plot assuming atoms are only in the $F = 1$ ground state. Assume we have some power broadening so that the width of the spectral features is 12 MHz. As always, don't worry about the amplitude of the spectral features.
(b) Reflect on your spectrum.
(c) Now assume you collected the spectrum but you used a crossover-free experimental setup. What does your spectrum look like now?

5.6 Answer the questions in Sect. 5.6.

References

1. Williams, W.D., Herd, M.T., Hawkins W.B.: Spectroscopic study of the $7p_{1/2}$ and $7p_{3/2}$ states in Cesium-133. Laser Phys. Lett. **15**(9), 095702 (2018). https://doi.org/10.1088/1612-202X/aac97
2. Maruko, C., Cölmek, N., Herd, M.T., Ahrendsen, K., Cabrales, B., Cannon, G., Davis, E., Guo, X., Karani, T., Wallace, A., Wisnauckas, K., Williams, W.D.: Spectroscopic study of the $4f^7 6s^2 \, ^8S^o_{7/2} - 4f^7 (^8S^o) \, 6\,s6p(^1P^o) \, ^8P_{5/2,7/2}$ transitions in neutral europium-151 and europium-153: absolute frequency and hyperfine structure. J. Opt. Soc. Am. B. **41**, 1217–1223 (2024). https://doi.org/10.1364/JOSAB.521181

Part II

Digging Deeper: Quantum Mechanics and Beyond

Quantum Mechanics vs. Classical Physics

6

Abstract

In this chapter, we explore the differences between quantum mechanics and classical physics, focusing on three main ideas: compatible and incompatible observables, states of an observable, and the superposition of states. By examining these concepts, you will gain insight into how measurements impact quantum mechanical systems and the ideas behind the uncertainty principle.

Learning Goals

By the end of this chapter, you should be able to understand:

- the importance of discrete energy states in quantum mechanics.
- compatible and incompatible observables.
- basis sets.
- superposition of states.

I'm often asked some variation of the question, "What is the difference between quantum mechanics and classical physics?" This is a great question! After teaching quantum mechanics many times and having numerous conversations with other atomic and nuclear physicists, I believe the best way to introduce the difference to new learners is to first understand two concepts: states and superposition. By grasping these two ideas, we can begin to explore why quantum mechanics is essential for explaining the world of the super small.

© The Author(s) 2025
W. Raven, *Atomic Physics for Everyone*,
https://doi.org/10.1007/978-3-031-69507-0_6

113

A note from Will: Many concepts discussed in this chapter are based on the orthodox or Copenhagen interpretation of Quantum Mechanics. Although it is the most popular interpretation, other interpretations exist, such as the many-worlds interpretation, de Broglie-Bohm pilot wave theory, and various collapse theories including Ghirardi-Rimini-Weber (GRW) theory and Continuous Spontaneous Localization (CSL). This remains an active area of exploration for many physicists and philosophers.

In addition, there are a few places where I simplify things a bit to make sure we focus on the main concepts. Finding the balance between 100% correct and still keeping things accessible was a challenge. For example, there are places where I will write a sum of functions where it should technically be an integral, but since calculus is not required for this book, I purposefully chose to leave them as sums to help us explore the concepts. For equations that should be integrals, I include a footnote to indicate that. For those who have not taken calculus, the ideas we discuss are still correct! We are just going to simplify the math a bit.

The other concept I simplify a bit is concerning compatible observables and sharing states. However, after discussions with my physics pedagogy friends, I decided to keep that language instead of exploring the delicate topic of projections and subspaces. For the advanced readers, keep this in mind. For new learners to quantum mechanics, this chapter is for you! If in the future you decide to learn more advanced quantum mechanics (and I hope you do!), please come back to this chapter to explore the subtleties of "commutation" and "subspaces."

6.1   What Is a State?

Definitions

- **Momentum:** A property of an object that is moving. For a classical object like a baseball, the formula for momentum is $p = mv$, where p is the object's momentum, m is the object's mass, and v is the object's velocity. An object's momentum will change if something acts on the object from the outside. The larger an object's momentum, the harder it is to stop. The unit of momentum is $kg \cdot m/s$.
- **Observable:** Something we can measure experimentally. Observables include, but are not limited to, energy, position, momentum, and angular momentum (angular momentum is the topic of Chaps. 7 and 8).

Physical systems, such as atoms, can exist in various conditions. They can be excited or relaxed, located in different places, moving or at rest, and so on. We describe these different conditions as "states," and use a mathematical "state function" to specify the exact state of a system at a given time.

As we talked about and explored in Part 1 of this book, atoms have *discrete* energy states.

> Two minute question: What does that mean?

Here is my answer: If we measure the energy of a quantum mechanical system, we will always find the system to have one value from a set of specific energies. The whole purpose of Part 1 of this book was how we experimentally find the energy between two of these discrete energy states. An electron in an atom can have energy E_1, E_2, E_3, etc. We will never find the electron with any other energy. We represent these discrete energy states in energy level diagrams throughout Part 1 of this book. Figure 6.1 is a copy of Fig. 5.12 as a helpful reminder. This is so important that I want to say it again: An electron in an atom can have energy E_1, E_2, E_3, etc. We will **never** measure the energy to be somewhere between E_1 and E_2.

Everything we can measure, which we call an observable, is represented by a state. The state of an observable is described by a mathematical function. Examples of things we can measure include energy, position, and momentum. As such, there are energy states, position states, and momentum states. Every energy we can measure is associated with some energy state. For example, suppose a system is in an energy state described by the state function ψ_1. If we measure the energy of that system, we will find the system has energy E_1. We will never measure any other value for energy. Often those states are discrete (like the energy of an electron in an atom) and sometimes they are continuous (like the position of an electron hanging

Fig. 6.1   A simplified energy level diagram for the transitions in cesium-133 near 455.6 nm

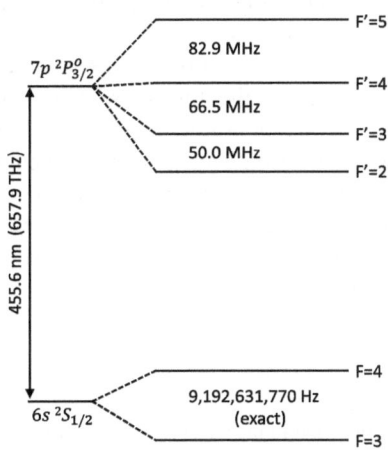

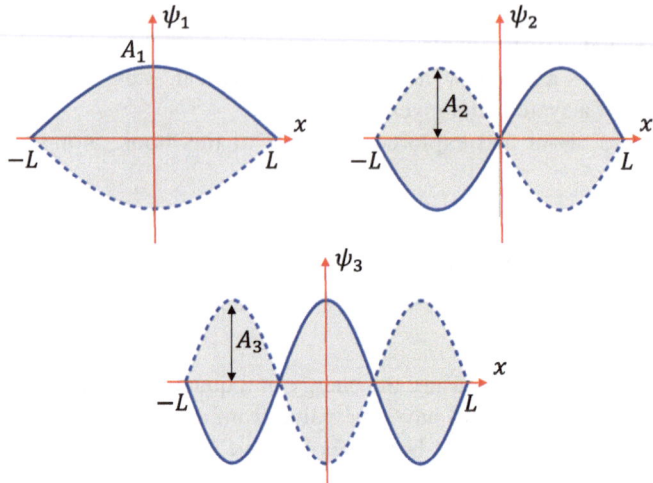

Fig. 6.2 The first three standing waves of a one dimensional rope or slinky. For this example, the rope is fixed at $x = -L$ and $x = +L$

out by itself in free space). *If we measure an observable, the quantum mechanical system will be in a state corresponding to that observable after the measurement.*

To help explore the idea of energy states, we will use a classical analogy that will serve as a helpful visualization tool. Imagine a string fixed at both ends, vibrating to create different standing waves, see Fig. 6.2. These are the same standing waves we discussed in Chap. 1 as a means of illustrating wave interference.

Now, let's shift our perspective from the classical realm to the quantum mechanical domain. If we were to treat this system quantum mechanically and measure its energy, we'd find discrete energy levels. These energy levels correspond to specific configurations of the standing wave—for instance, one loop, two loops, three loops, and so on as shown in Fig. 6.2. Unlike in the classical scenario where the wave can have any arbitrary energy, in the quantum realm, the energy is constrained to certain discrete values.

To formalize this, we use mathematical expressions known as wavefunctions,[1] denoted by the lowercase Greek letter psi (ψ). Each wavefunction, labeled as ψ_n, corresponds to a particular energy level, which will have energy E_n. The wavefunction encapsulates the behavior and properties of the system at that energy level. In quantum mechanics, the equation we use to calculate the wavefunctions and energies is called the **Schrödinger equation**. The Schrödinger equation is a partial differential equation and solving it for a system provides not just the spatial form of the wavefunction, but also how they evolve in time. If you take a quantum

---

[1] In quantum mechanics, we prefer the phrase wavefunction over state function.

mechanics class, you will spend a good amount of time solving the Schrödinger equation for different physical scenarios.

Returning to our example of the vibrating quantum mechanical string, we can now describe the wavefunctions for specific energy states. The wavefunction ψ_1 corresponds to the state with one loop of energy, denoted by E_1. Its functional form, see Fig. 6.2, is described by $\psi_1 = A_1 \cos\left(\frac{\pi x}{2L}\right)$, where A_1 represents the amplitude and $2L$ denotes the length of the string. Similarly, ψ_2 represents the state with two loops and has a functional form of $\psi_2 = A_2 \sin\left(\frac{\pi x}{L}\right)$.

Summary So Far

If a quantum mechanical system has a particular energy, we say the system is in an energy state. Mathematically, that energy state is described by the wavefunction ψ_n. If a quantum mechanical system is in an energy state described by the wavefunction ψ_n and we measure its energy, we will measure the system's energy to be E_n. We will never measure anything else.

For the example in this section, n represents how many loops of energy the system has, so ψ_3 is the third picture in Fig. 6.2. In bra-ket notation, see Sect. 3.6, we would write either $|\psi_n\rangle$ or $|n\rangle$. There are other quantities (obersvables) that we can measure, and each thing we measure can change the shape of the wavefunction. Figure 6.3 shows an example of three position states. If we measure the position of a quantum mechanical particle, we would find the particle somewhere between the start and end of the string. Mathematically, we label those position states ψ with a subscript, like ψ_x. Notice that the position wavefunction is zero everywhere

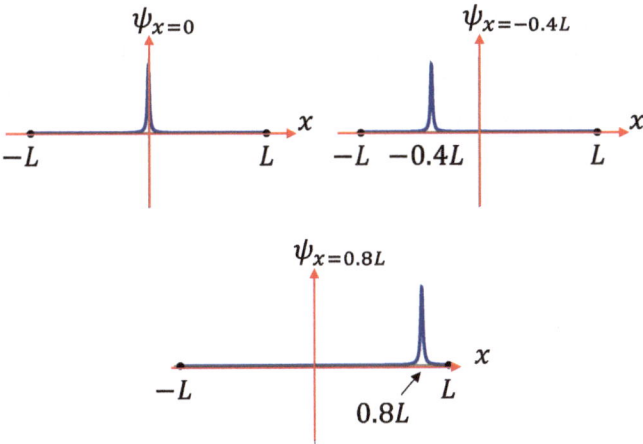

Fig. 6.3 Examples of 3 different position states. Each position state has a specific position associated with it

except for where the particle was measured. For the first example, the particle was measured to have a position at $x = 0$. For the second example, the particle was measured to have a position at $x = -0.4\ L$.

> **Super Important**
> For this quantum mechanical system, the energy states ***are not*** position states. They look completely different.

If we measure the energy of a quantum mechanical system, the system will be in an energy state, for example one of the energy states shown in Fig. 6.2. If we measure the position of a quantum mechanical system, the system will be in a position state, for example one of the position states shown in Fig. 6.3. *These are not the same states!!*

That last sentence is super important. If I told you a quantum mechanical system had energy E_3, that statement comes with the understanding that our system is in an energy state (ψ_3 or $|3\rangle$) that has energy E_3.[2] If the system was in the energy state ψ_3 and I were to ask, "Where exactly is the quantum mechanical particle?", we wouldn't be able to answer that question. The wavefunction, see the graph with 3 loops in Fig. 6.2, is spread out over the entire range from $x = -L$ to $x = +L$ and this is *not* a position state, see Fig. 6.3. While we can't state precisely where the particle is, we can, as we will see in Sect. 6.3, answer the question, "What are the probabilities of finding the particle in various locations?"

Likewise, if I told you a quantum mechanical particle was at a position $x = 0.1$ nm, that statement comes with the understanding that our system is in a position state with a location at $x = 0.1$ nm. If I were to ask, "What is the energy of the quantum mechanical particle?", we wouldn't be able to answer that question exactly because the system is not in an energy state. Like with the example given above where we know the energy but not the exact position, we will be able to answer the question, "If the particle is in the position state $\psi_{x=0.1\ \mathrm{nm}}$, what are the probabilities of finding the particle with a particular energy?"

---

[2] As we work through the next few chapters, we will add more parameters to describe the state. Eventually, the state will look something like $|n\ \ell\ s\ j\ m_j\rangle$. The 3 in this example is the n.

Stepping Back

Take a look at one of the energy states shown in Fig. 6.2 and ask the question, "Where is the particle?" The best you can answer is, "The wavefunction is spread out from $x = -L$ to $x = L$. It is not at any single location." The energy has a single value, but the exact position is unknown.

Far less intuitive is if we repeated this thought experiment with position states. Take a look at one of the position states shown in Fig. 6.3 and ask the question, "What is the energy of the particle?" Our intuition tells us there is an answer to this question. However, in quantum mechanics, a particle only has a specific energy if it is in an energy state. A position state is *not* an energy state. The best you can answer is, "The energy is spread out. It does not have a single value."

Take as much time as you need to try and understand that last paragraph. That last paragraph is very counterintuitive to our everyday experience. In classical physics, if a car is traveling down the road, I can tell you its position and energy. In quantum mechanics, we cannot know both. If the energy is well defined, it is in an energy state. If the position is well defined, it is in a position state.

There are two really important things to take away from this section:

- If we measure the energy of a quantum mechanical system, that system will be in an energy state with a specific energy. If we measure the position of a quantum mechanical system, that system will be in a position state with a specific position.
- Energy states and position states are not the same. They look completely different. If we measure the energy of a system, the system will now be in an energy state and the position of the particle is spread out. Similarly, if we measure the position of a system, the system will now be in an position state and the energy of the particle is spread out.

6.2 Compatible vs. Incompatible Observables and the Uncertainty Principle

---

Definitions

- **Incompatible observables:** If two observables cannot be precisely measured at the same time, they are called incompatible. Measuring one observable changes the system, making subsequent measurements of the other observable unpredictable.
- **Compatible observables:** If two observables can be precisely measured at the same time, they are called compatible. Measuring one observable does not change the system in a way that affects the measurement of the other observable. Remeasuring either observable will return the same value as initially measured.

We have learned that for quantum mechanical systems, energy and position are *incompatible observables*. If the system is in a position state, it is not in an energy state and vice versa. *If we measure the position of the system, it is now in a position state. Since that is not an energy state, we cannot predict the exact outcome when we measure the system's energy.*

If there are two observables that share the same set of states, we say that they are *compatible observables*. For the system in Sect. 6.1, there are, unfortunately, no compatible observables, but we will be able to explore compatible observables using systems in Chap. 7. So, for now, let's just suppose there is some other observable that is compatible with energy. Let's call that observable A, described by the wavefunction $\psi_{A,n}$ with possible measured values A_n. If we measure the energy of a quantum mechanical system, the quantum mechanical system will now be in an energy state. Let's say we measured the energy of the system to be E_3, which is the third picture with three loops in Fig. 6.2. Now we measure A and obtain A_3. If we went back and measured energy, we will find that the system is still in the state with three loops with energy E_3. Next, we measure A. If energy and observable A are compatible, we will measure A_3. This is because two compatible observables share a set of states. In other words, if the energy state has the wavefunction with three loops (ψ_3), and we measure A, we will find $A = A_3$ and that the wavefunction is unchanged. It still has three loops.

The Uncertainty Principle Part 1
The compatibility or incompatibility of two observables is summarized by the uncertainty principle. If two observables are compatible, they share a set of states. We can measure both observables repeatedly and obtain consistent values. If the two observables are incompatible, they do not share a set of states. Measuring one observable disrupts the outcome of the measurement of the other.

As you see, states are extremely important in quantum mechanics. Every observable has a set of associated states. Each energy state corresponds to a specific energy. If we measure the system's energy, it will be in one of those states with the associated energy. Similarly, measuring position, momentum, or any other observable places the system in one of the states for that observable. Some observables share a set of states (they are compatible, and subsequent measurements do not alter the previous values), while others have different states (they are incompatible, and knowing one means we don't know the other).

6.3 Superposition of States

Definitions

- **Superposition:** A concept in math and physics that says a function can be constructed as the sum of two or more other functions.
- **Basis set:** All of the states for a particular observable for a particular system. All of the possible energy states (for example, 1 loop, 2 loops, 3 loops, 4 loops, etc. for the vibrating quantum mechanical string shown in Fig. 6.2) make up the energy basis set for that system. Similarly, all possible position states (Fig. 6.3 shows a few position states for the vibrating quantum mechanical string) make up the position basis set.

Another really important concept to understand is superposition. Let's explore superposition through an example. Suppose we measure the position of a quantum mechanical system and find it in the position state shown in the top right picture of Fig. 6.3. Although this is not an energy state, we can still infer some information about the system's energy. The concept of superposition will allow us to determine *the probability* that, upon measuring energy, the system will have one loop, two loops, three loops, etc.

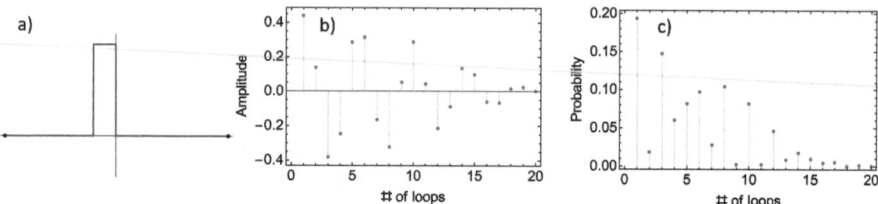

Fig. 6.4 (**a**) The state for our thought experiment. (**b**) The amplitude of each energy state needed to construct the state. (**c**) The probability that, upon measurement of energy, we find the system in a particular energy state

What we will do is "build" a position state by cleverly adding together all of the energy states. Using more mathematical language, we will construct a position state from a superposition of states from the energy basis set. Although this sounds strange–since energy states have loops and the position state we are interested in is a spike–we can, remarkably, build a spike from loops!

To help understand this concept, let's do a thought experiment where we put the system into a state that isn't a position state, but is mathematically simpler, in order to explore the concept of superposition. At the end of this section, I will show graphs to build the position state seen in the top right of Fig. 6.3 from the energy basis set. The example state we will use to explore superposition is shown in Fig. 6.4a.

Mathematically, the energy states (those shown in Fig. 6.2 plus all the other possible energy states) can be added together to construct the desired state. More formally, we say that a state can be constructed as a superposition of the energy basis set:

$$\psi = A_1\psi_1 + A_2\psi_2 + A_3\psi_3 + \dots = \sum_{i=1}^{\infty} A_i\psi_i \qquad (6.1)$$

where A_i is the amplitude of ψ_i that is mathematically determined to reconstruct the desired state ψ. The set of all states ψ_i is called the energy basis set. This equation represents the concept of superposition. We are constructing a specific state from a superposition of states from the energy basis set. The math required to calculate the amplitudes A_i is complicated and requires integrals, so I am just going to show you the results of the math in Fig. 6.4b. In practical terms, we start with the energy state with 1 loop (left picture in Fig. 6.2) and set the amplitude to 0.44. Next, we take the energy state with 2 loops (middle picture in Fig. 6.2), set the amplitude to 0.14, and add it to the energy state with 1 loop that had an amplitude of 0.44. We then take the energy state with 3 loops, set the amplitude to −0.38, and add it to the first two energy states. This process repeats, and the more energy states we include, the closer we get to the actual state. This is illustrated in Fig. 6.5. Adding the first two energy states together (top row) with the appropriate amplitude doesn't resemble the desired state, but after adding the first 30 energy states (third row), the resultant graph starts to look like the desired state. A superposition of the first

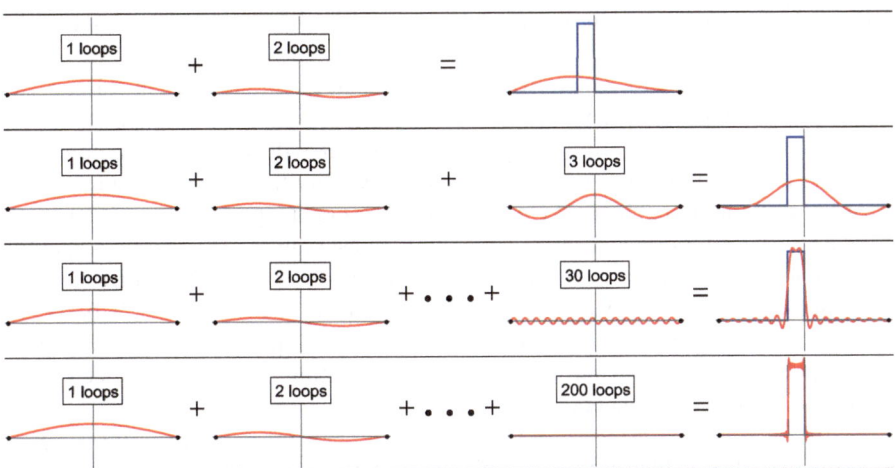

Fig. 6.5 An illustrative example of adding more and more energy states together to build the desired state. The amplitude of the energy state with 1 loop is 0.44, the amplitude of the energy state with 2 loops is 0.14, 3 loops has an amplitude of −0.38, etc. For the bottom row, the energy state with 200 loops looks flat to our eye, but it has a very small amplitude of 0.019

200 energy states (bottom row) results in a close reconstruction. Remarkably, loops can construct this state! Even more astonishingly, you can create any state you want (as long as it is a well-defined single-valued function) from the energy basis set. In other words, you can construct any state from a superposition of energy states.

Since we can construct a state out of energy states, we say that the state is a superposition of states from the energy basis set. Similarly, we could construct an energy state from a superposition of states from the position basis set. All we need is some mathematical method to determine the amplitudes of each of the position states so that when we add them all together we get the desired energy state.

In quantum mechanics, this is a very general idea. We can always construct a single state from any basis set using a superposition of states from another basis set. For example, we can construct a momentum state from a superposition of states from the position basis set, a position state from a superposition of states from the momentum basis set, a momentum state from a superposition of states from the energy basis set, and so on. This is a neat math trick, but is it useful? *It is perhaps the most useful math trick in all of quantum mechanics.*

Suppose we are in the state given by Fig. 6.4a. Now we want to measure the energy of the system. The amplitudes given in Fig. 6.4b can be used to calculate the probability that, upon measurement of energy, we will find the system in a particular energy state. All we need to do is square that amplitude. For example, the amplitude of the first energy state (1 loop) is 0.44. The probability that, upon measurement of energy, we find the system with one loop of energy is $0.44^2 = 0.19$, or 19%. The amplitude of the third energy state (3 loops) is −0.38. The probability that, upon measurement of energy, we find the system with 3 loops of energy is $(−0.38)^2 =$

0.15, or 15%. The probabilities are given in Fig. 6.4c. This rule is universal. If the system is in a particular energy state and we want to measure position, there is an associated probability for where we would find that quantum mechanical particle. Those probabilities are found by first writing the energy state as a superposition of all the position states from the position basis set with the correct amplitudes. Squaring those amplitudes[3] will tell you the probability that, upon measurement of position, a quantum mechanical particle will be found at a particular position.[4]

The above discussion is for two incompatible observables. The two incompatible observables have distinct basis sets, one for each observable. Measuring one of the observables puts the system in a single state from that observable's basis set. However, that single state can be constructed from a superposition of states from the other basis set. But, what if we had two compatible observables? Let's recap some of the important concepts:

• If the two observables are compatible, they share a basis set.
• If we measure one observable, the system is now in a state of that observable.
• If the two observables are compatible, measuring one observable does not change the system in a way that affects the measurement of the other observable.

Interestingly, we can use superposition to answer this question. Suppose energy and our made up observable A are compatible observables. We measure the energy of the system and find the system has energy E_3. The system is now in the energy state ψ_3, which is also the same as $\psi_{A,3}$. Let's write that energy state as a superposition of states from the basis set for observable A:

$$\psi_3 = 0 * \psi_{A,1} + 0 * \psi_{A,2} + 1 * \psi_{A,3} + 0 * \psi_{A,4} + ... \tag{6.2}$$

For the system we are exploring, building the energy state from a superposition of A states is really easy! ψ_3 and $\psi_{A,3}$ are literally the same sate. The probability that, upon measurement of A, we find the system with A_1 (the value associated with $\psi_{A,1}$) is 0%. The probability that, upon measurement of A, we find the system with value A_3 (the value associated with $\psi_{A,3}$) is 100%. The system is already in that state! Constructing the energy state from the A basis set is easy because there is one state from the A basis set that perfectly matches the energy state.

---

[3] For completeness, the amplitudes could be complex numbers, which is something beyond the scope of this book. In the future, if you see an amplitude that is complex, you calculate the modulus squared of the amplitude.

[4] For completeness (again ☺), position and momentum are, mathematically, a bit harder to deal with since their measured values are continuous and not discrete like energy. The idea we explored is the same, but instead of the summation in Eq. 6.1 we would have an integral. As such, we would state something like, "There is a 25% probability we would find the particle between $x = 0.100$ nm and $x = 0.102$ nm". This is a minor detail, but one I wanted to include a footnote for those who have taken some more advanced math.

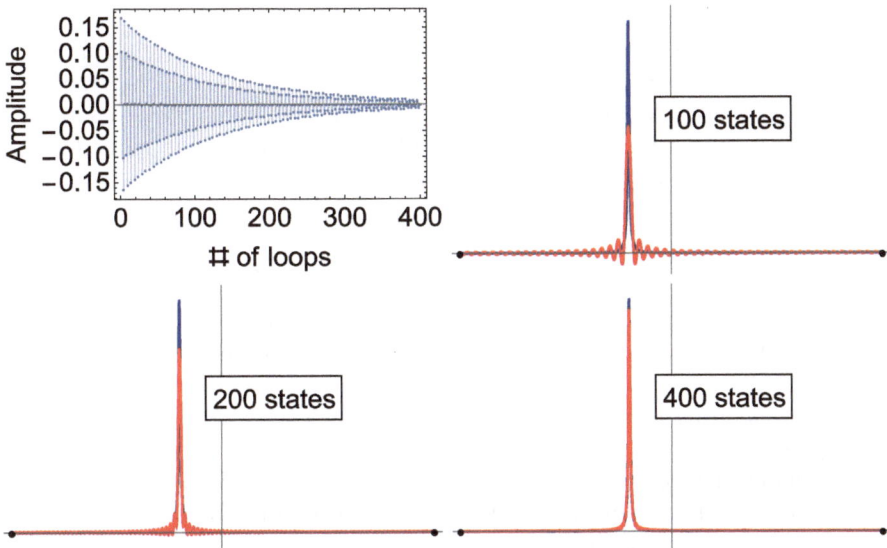

Fig. 6.6 Constructing the position state from the middle picture in Figure 6.3. We need over 400 energy states to build this one state! Notice the position state in blue is noticeably taller than the reconstructed states even when we use the first 200 energy states. The position state is still a bit taller even with 400 energy states!

Now that we have explored the concept of superposition, let's build a more realistic position state, similar to the top right graph in Fig. 6.3, from the energy states. Figure 6.6 shows the mathematical results needed to construct that position state from the energy basis set. As you can see, we needed at least twice as many energy states to get a good approximation for a position state. Despite needing many more energy states to accurately reconstruct a single position state, the result remains the same: if the quantum mechanical system were in that position state and we measured its energy, we could get many different answers with probabilities given by the squared amplitudes.

The Uncertainty Principle Part 2
Measuring an observable puts the system into a single state for that observable and in a superposition of states for a second observable. If a second observable shares states with the basis set of the first observable, the value of that second observable is already determined. We can measure both quantities as many times as we'd like, and we will never change the state of the system. We will always get the same answer as the first time we measured.

(continued)

If the two observables have different basis sets, knowing the value of one observable does not inform you of the value of the other. These observables are incompatible. Measuring one observable places the system in a superposition of states for the other observable where none of the amplitudes are 1. Therefore, the state of the second observable after measurement is probabilistic.

States and superposition of states are core concepts in quantum mechanics. Let's summarize the entire chapter using two generalized observables α and β.

- Both α and β have states associated with them, denoted as $\psi_{\alpha,1}$, $\psi_{\alpha,2}$, ... and $\psi_{\beta,1}$, $\psi_{\beta,2}$, ...
- The set of states that are needed to describe all possible values for α is called the basis set of α. The set of states that are needed to describe all possible values for β is called the basis set of β.
- Each state has a measurable value associated with it. If the system is in state $\psi_{\alpha,7}$ and we measure α, we will get the value associated with that state, α_7. We will never measure, for example, α_5.
- If α and β are incompatible, they have different basis sets. If they are compatible, they share a basis set.
- A state for α can be written as a superposition of states for β, and vise-versa. For example,

$$\psi_{\alpha,7} = A_{\beta,1}\psi_{\beta,1} + A_{\beta,2}\psi_{\beta,2} + A_{\beta,3}\psi_{\beta,3} + ... \tag{6.3}$$

or

$$\psi_{\beta,137} = A_{\alpha,1}\psi_{\alpha,1} + A_{\alpha,2}\psi_{\alpha,3} + A_{\alpha,3}\psi_{\alpha,3} + ... \tag{6.4}$$

- The square of the amplitudes in the above equations tell us the probability, upon measurement, we will find the system in that state with the value associated with that state. For example, if the system was in state $\psi_{\beta,137}$ with value β_{137} and we measured α, the probability the system is now in a particular state of α, say state $\psi_{\alpha,3}$ with value α_3, is the amplitude squared, $A_{\alpha,3}^2$.
- Suppose we measured β and got β_{137}. Next, let's say we measure α and get α_7. If α and β are incompatible and we then re-measure β, it is highly likely we will get something other than β_{137}.

6.4   The Energy Basis Set for a Quantum Harmonic Oscillator

The quantum harmonic oscillator is one of the first quantum systems learners encounter in a quantum mechanics class, and it is a very practical system to study. It can be used to model many physical systems including molecular vibrations, see Fig. 6.7. Consider two atoms connected by chemical bonds to form a single molecule. A classical model for this system is to have the two atoms connected by a spring. If the two atoms are not moving and the spring is not stretched or compressed, the system would just sit there at rest and not vibrate. However, if the spring is stretched slightly, it tries to pull the atoms back closer together. As the atoms move closer, the spring passes its equilibrium length and compresses. Once compressed, the spring tries to push the atoms apart until it stretches past equilibrium again, and the process repeats. This is an oscillation. In a quantum mechanical version, this system will have discrete energy levels, meaning it will have specific energy states. If the spring from this classical analogy were behaving quantum mechanically, the spring would only oscillate at specific frequencies.

We can use the Schrödinger equation to find the energy states and their energies for the quantum harmonic oscillator. The energy states with the four smallest energies for the quantum harmonic oscillator are shown in Fig. 6.8. Notice the similarities with the standing waves on a string example in Sect. 6.1. Although the shape of the states look different, the lowest energy state has 1 loop, the second lowest energy state has 2 loops, etc. If we measure the vibrational energy of a molecule modeled by the quantum harmonic oscillator, we would find the system in one of the energy states such as those shown in Fig. 6.8. Each state has a defined, discrete energy. All of the same bullet points that summarized Sect. 6.3 are still true! Observables that are compatible with energy, for example the frequency of vibration, have the same basis set, allowing us to measure all

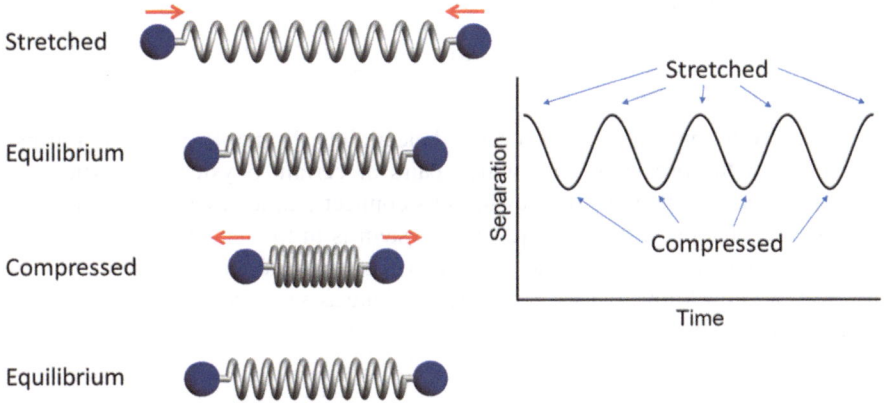

Fig. 6.7 Left: Two atoms connected by a spring. This is a model for two atoms connected by molecular bonds. Right: The separation of the atoms as a function of time

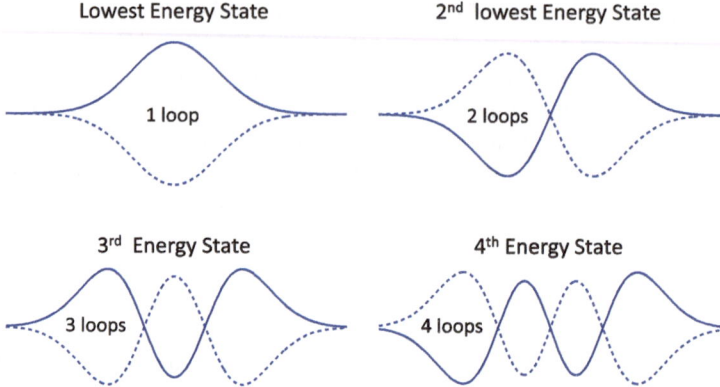

Fig. 6.8 The energy states with the four smallest energies for the quantum harmonic oscillator. Each of these energy states also corresponds to a specific energy or frequency of vibration of the atoms in the molecule

compatible observables repeatedly without disturbing the outcome of measuring other compatible observables. Observables that are incompatible with energy have different basis sets. If we measure energy, the system will be in a state from the energy basis set and a superposition of states for the incompatible observable's basis set. The amplitudes in the superposition formula allow us to calculate the probability of what we will measure for the incompatible observable.

6.5   The Uncertainty Principle Part 3

Reminder Definition

- **Reduced Planck's constant (h-bar):** $\hbar = \frac{h}{2\pi} = 1.054 \times 10^{-34}\,\mathrm{kg \cdot m^2/s}$;

The most common explanation of the Uncertainty Principle is about intrinsic limitations to the precision with which pairs of certain physical properties of a particle can be simultaneously known. Let's connect that idea with what we talked about in this chapter. Suppose a quantum system is in the rectangular state shown in Fig. 6.4a. This is neither an energy state, a position state, nor a momentum state. However, we now know that we can write this state as a superposition of states from any observable's basis set.

For momentum, we would write[5]

$$\psi = A_{p,1}\psi_{p,1} + A_{p,2}\psi_{p,2} + A_{p,3}\psi_{p,3} + \ldots = \sum_{i=1}^{\infty} A_{p,i}\psi_{p,i}$$

where $\psi_{p,i}$ are the states making up the momentum basis set. Similarly, we could write the state as a superposition of states from the position basis set:

$$\psi = A_{x,1}\psi_{x,1} + A_{x,2}\psi_{x,2} + A_{x,3}\psi_{x,3} + \ldots = \sum_{i=1}^{\infty} A_{x,i}\psi_{x,i}$$

where $\psi_{x,i}$ are the states making up the position basis set. If the system was in the state shown in Fig. 6.4a, and we measured momentum, we would get a *range* of possible values. More specifically, we have a $A_{p,1}^2$ chance of measuring p_1, a $A_{p,2}^2$ chance of measuring p_2, and so on. Equivalently, we would have some average value of momentum. For example, if we had a 100% chance of measuring p_4, the range (or spread) of possible measurements for momentum is 0 while the average value would be p_4. Let's call the range Δp and the average value we measure $\langle p \rangle$. Similarly, if we instead had measured position, we would get a range of possible position measurements, which we will call Δx, and an average position measurement, which we will call $\langle x \rangle$.

If two observables A and B are compatible, there is a scenario where $(\Delta A)(\Delta B) = 0$. That scenario would be when the system is in a state of A, which also happens to be a state of B since they are compatible. In that scenario, there is no range of measurements for either A or B: $\Delta A = 0$ and $\Delta B = 0$. However, if the system was not in a state of A and B, there would be a range of possible measurements for both observables. Therefore,

$$(\Delta A)(\Delta B) \geq 0 \tag{6.5}$$

The uncertainty principle for two compatible observables.

This equation covers all possible scenarios for two compatible observables.

Now let's explore incompatible observables, for example momentum and position. Since momentum and position are incompatible, then there is no situation where $(\Delta x)(\Delta p) = 0$. If we did the math to determine the relationship, we would find that there is a minimum possible value to this product. For these two observables, the minimum possible value is

$$(\Delta x)(\Delta p) \geq \frac{\hbar}{2} \tag{6.6}$$

---

[5] For the advanced reader, the next two equations should technically be integrals since momentum and position have a continuous range of possible values, unlike energy that has discrete values.

No matter what state you consider, there will always be a range of possible measurements for at least one of the observables. You might be asking yourself the question, "But wait, if the system was in a position state, wouldn't $\Delta x = 0$ and thus $(\Delta x)(\Delta p) = 0 \times (\Delta p) = 0$?" If the system was in a position state, the range of possible measurements for momentum would be ∞. So, we would have $(0)(\infty)$, which is undefined. However, if we were to carefully take the limit as $\Delta x \to 0$ and $\Delta p \to \infty$, the product is still greater than $\hbar/2$.

For every pair of incompatible observables, the product of the ranges is greater than some minimum value. The result is easier to read for some pairs of incompatible observables than others. For example, Eq. 6.6 is fairly straightforward: the range of possible measurements of position multiplied by the range of possible measurements of momentum is always greater than or equal to the constant $\hbar/2$. Other times, the uncertainty principle is harder to interpret. The uncertainty principle for position and energy is

$$(\Delta x)(\Delta E) \geq \frac{\hbar}{2m}|\langle p \rangle| \qquad (6.7)$$

This equation is a bit harder to read. I would interpret it as follows: for a given state, the product of the spread of possible position measurements and the spread of possible energy measurements will always be greater than a constant times the magnitude of the average value of possible momentum measurements. This is a formula that you must be careful with. There are scenarios where $|\langle p \rangle| = 0$, but that does not mean that there is a scenario where position and energy are compatible. This just says that the product of their ranges must be greater than or equal to 0.

6.6   Looking Forward

This chapter solidifies some concepts we explored in the first part of this book while adding some more. Some observables are compatible (i.e., they share a basis set and subsequent measurements do not disrupt the values of the observables) and some observables are incompatible (i.e., they do not share a basis set and measuring one observable puts the system in a superposition of states for the other observable). The next questions to think about are

- What observables are compatible with energy?
- What observables are incompatible with energy?

We will explore these questions in the next few chapters.

A reflection from Will: This was a really hard chapter for me to write!^{a} There are so many fun things to explore in quantum mechanics that I had a difficult time narrowing it down to just a few important concepts. So, please take this chapter as an initial introduction. If you choose to go on and learn more about quantum mechanics (and I hope you do!), you can use this chapter as a base to build more understanding. There are many fascinating things to explore like quantum entanglement, Bell's Inequality, Ehrenfest's theorem, and the no-cloning theorem, just to name a few. The uncertainty principle also has much more to explore and learn. As with all things worth learning in life,

$$\text{questions} + \text{repetition} + \text{critical thinking} = \text{mastery}$$

---

^{a}And I learned a lot by doing so.

Problems

6.1 In your own words, summarize the Uncertainty Principle.

6.2 Suppose a quantum system has two incompatible observables. We now know two important principles:

1. A state for one of the observables can be constructed from a superposition of states from the basis set of the other observable.
2. The probability that, upon measurement of the other observable, we find the system in a particular state can be calculated by squaring the amplitude of that state.
   (a) What would we get if we squared all the amplitudes and added them up? In other words, what is:

$$\sum_{i=1}^{\infty} A_i^2 =?$$ (6.8)

   Explain.
   (b) Now suppose there is a quantum system with two compatible observables. Do the two principles listed in the paragraph above still hold true for two compatible observables? Why or why not?

6.3 For the quantum harmonic oscillator, are position and energy compatible observables? Explain.

6.4 Suppose we had a quantum mechanical system in the energy state represented by the second state in Fig. 6.2 (the energy state with two loops).

(a) Draw that energy state on a piece of paper. Reconstruct that energy state by drawing position states (Fig. 6.3) so that when you add them together you get that energy state.
(b) Now you want to measure position. Where are you most likely to find the quantum mechanical particle?
(c) Let's put the system back into that energy state. What is the probability, upon measurement of position, that the particle is found exactly in the middle? Why?

6.5 (The Measurement Game)   Suppose we have three observables: A, B, and C. A and B are compatible observables while C is incompatible with both A and B. Below is a list of measurements that happen chronologically. If you will measure a specific value, state that value you will measure. For example, if you know that you are going to measure B_4, state, "I will measure B_4 with 100% probability." If you will not measure a specific value, state, "I can measure a variety of things." and then make one up, for example, "I measured A and got A_{17}."

You measure A and get A_4.
Now you measure B. What can/will you measure?
Now you measure A. What can/will you measure?
Now you measure C. What can/will you measure?
Now you measure A. What can/will you measure?
Now you measure C. What can/will you measure?
Now you measure B. What can/will you measure?
Now you measure A. What can/will you measure?
Now you measure B. What can/will you measure?
Now you measure A. What can/will you measure?
Now you measure A. What can/will you measure?
Now you measure C. What can/will you measure?

6.6 Write Eq. 6.1 using bra-ket notation.

6.7 Just for fun, here is a crossword with some terms from this chapter. For each word, write a clue.

Word list:
Basis
Classical
Energy
Momentum
Observable
Quantum
States
Superposition

Angular Momentum

7

Abstract

In this chapter, we explore angular momentum, a key concept in quantum mechanics and atomic physics. We cover the quantization of angular momentum, the role of quantum numbers, and their impact on atomic states. We will learn about the different types of angular momentum, the states representing these types of angular momentum, and how angular momenta are added together in quantum mechanics. The chapter also examines the compatibility of different observables, such as the magnitude and components of angular momentum, and the impact of measurements on quantum states.

Learning Goals

By the end of this chapter, you should be able to understand:

- that angular momentum is a vector that has both magnitude and direction.
- that quantum mechanics restricts angular momentum to certain discrete magnitudes and directions.
- that there are multiple angular momentum vectors in an atom, and understanding how these vectors add together impacts the energy of a particular state.
- that quantum numbers are key tools for understanding angular momentum in quantum mechanics.
- the concept of compatible and incompatible observables in the context of angular momentum measurements.

© The Author(s) 2025
W. Raven, *Atomic Physics for Everyone*,
https://doi.org/10.1007/978-3-031-69507-0_7

7.1   Definitions

Note There are *a lot* of new definitions in this section. We will talk about and try to understand all of them, but don't worry too much about memorizing everything right now. We will use some of the definitions continually throughout the rest of the book. In future discussions, you should come back to these definitions as needed. We are all learners, and learning takes repetition.☺

Momentum This is the same definition from Chap. 6: a property of an object that is moving. For a classical object like a baseball, the formula for momentum is $p = mv$, where p is the object's momentum, m is the object's mass, and v is the object's velocity. An object's momentum will change if something acts on the object from the outside. The larger an object's momentum, the harder it is to stop. The unit of momentum is kg m/s.

Force Below are two definitions of force. The first is the most common definition. However, it is only true if the object's mass is not changing, which is a very common scenario. The second is the real definition of force. The unit of force is $kg\,m/s^2$, which we also call a newton (N).

Definition #1 An external interaction that causes an object to accelerate (i.e., change velocity). Imagine pushing a block up a ramp. You are pushing on the block and causing its velocity to change. That push is an external interaction (force) on the block. Any single force acting on an object will cause the object to accelerate (change velocity). There can be multiple forces acting on an object. Sometimes those forces cancel each other out. If there are multiple forces that perfectly balance one another so the net force is 0, then the object will not accelerate. However, if the net force is not 0, it will accelerate. This is summarized by Newton's 2nd Law: "The sum of all forces acting on an object causes an object with mass to accelerate."

Definition #2 An external interaction that causes an object's momentum to change. Newton's 2nd Law for momentum: "The sum of all forces acting on an object causes an object's momentum to change."

Torque A measure of how effective a force is in causing an object to rotate. The two things that matter for torque are the size of the force that causes the rotation and how far from the axis of rotation that force is being applied. Imagine you have a bicycle wheel. If you exert a force horizontally at the edge of the wheel, the wheel starts to spin. Since that force caused a rotation, there is a torque. Applying that same force on the axle or pushing on the tire directly towards the axle does not cause any rotation, so in those two scenarios there is no torque. The unit for torque is $Nm = kg\,m^2/s^2$. Interestingly, torque and energy share the same unit: $kg\,m^2/s^2$. However, torque is not energy; they are incredibly different!! To avoid confusion, we tend to not write the unit $kg\,m^2/s^2$ for either energy or torque. We use joules (J) for energy and newton meters (Nm) for torque. Torque will be important in this chapter.

Angular Momentum A property of an object that is rotating. It is the thing that changes when a torque is applied to the object. There is a parallel between the ideas of force and momentum, and torque and angular momentum. Momentum is a property of an object that is moving along a straight line, and it changes when a force is applied to the object. The larger an object's momentum, the harder it is to stop the object from moving (a bigger force is needed). In the same way, angular momentum is a property of an object that is rotating, and the object will start to rotate faster or slower when a torque is applied (a bigger torque makes the change happen faster). The unit of angular momentum is $kg\,m^2/s$. Angular momentum is *extremely* important in quantum mechanics and atomic physics.

Reduced Planck's Constant (h-Bar) $\hbar = \frac{h}{2\pi} = 1.054 \times 10^{-34}\,kg\,m^2/s$.

▶ **Important Comment** Both Planck's constant and the reduced Planck's constant have units of angular momentum, $kg\,m^2/s$, which can also be written as joules times seconds, Js. The reduced Planck's constant will be used a lot in this chapter. You will see why we call it a fundamental constant.

Quantum Number In quantum mechanics, a lot of different properties are quantized, i.e., can only have certain values. We already know that the energy levels of an atom are quantized. Other properties, like angular momentum, are also quantized. When something is quantized, a quantum number is usually associated with that property. A quantum number is an integer or half-integer. Knowing a quantum number will allow someone to calculate some property of an atom. Quantum numbers have no units.

As an example, we will soon talk about the electronic orbital angular momentum quantum number (that is a mouthful!). That quantum number is represented by a script ℓ. ℓ is a quantum number that can be zero or a positive integer: 0, 1, 2, 3, etc. ℓ is a discrete number because, according to quantum mechanics, the size or magnitude of the orbital angular momentum of electron is found to have only certain, discrete values. If I tell you that a state in an atom has $\ell = 2$, we can calculate the magnitude of the electron's orbital angular momentum. $\ell = 2$ is not the actual size of the orbital angular momentum, but the quantum number can be used to calculate it from the formula $\sqrt{\ell(\ell + 1)}\hbar$. Notice that everything in that formula is either the quantum number ℓ or the reduced Planck's constant. In a way, knowing a quantum number is very similar to how spectroscopists use cm^{-1} as an energy unit. cm^{-1} is not the correct unit for energy, but we can multiply it by fundamental constants to get the actual energy. The same is true with quantum numbers. Knowing a quantum number is equivalent to knowing a specific property. You might have to do some math, but the only other parameters in the equation should be fundamental constants. So, being told $\ell = 2$ is the same thing as being told that the magnitude of the electron's orbital angular momentum is $\sqrt{6}\hbar = 2.582 \times 10^{-34}$ Js. We will soon find that we need many quantum numbers to describe a particular state in an atom.

7.2   Angular Momentum

Electrons, protons, and neutrons all have angular momentum. A classical analogy is thinking about the moon orbiting the earth.[1] The moon has orbital angular momentum because it is orbiting the earth. The moon is also spinning on its own axis, so it also has "spin" angular momentum. So, the moon has two types of angular momentum: orbital and spin. Likewise, an electron in an atom can have orbital angular momentum and spin, which is sometimes called intrinsic angular momentum. Interestingly, an electron always has intrinsic angular momentum, but it does not always have orbital angular momentum. One of the craziest things about quantum mechanics is that we don't have a great analogy to think about the orbital angular momentum or spin of the electron. The electron is not orbiting around the nucleus or spinning on its axis like the moon, but it has the same properties as if it were. The electron, as far as we can tell, has no size! So, how can it be spinning? Imagine how confusing that must have been when physicists were first trying to understand the electron.[2] It has all the properties one would expect for a ball spinning on its axis, but it is not a ball and it is not spinning! Understanding electron spin is still a wonderful mystery.

Since there are two types of angular momentum, we could ask the question, "What is the total electronic angular momentum of the electron in an atom?" The total electronic angular momentum is not a simple sum of orbital angular momentum and spin. In other words, you cannot just add the two angular momenta together like $2 + 3 = 5$. Angular momentum is something called a vector, which is something with magnitude (i.e., size or length) and direction. To get the total electronic angular momentum you have to add the orbital angular momentum and spin together in their vector forms. As a classical analogy, imagine you walk 30 m due north, see Fig. 7.1. That is a vector because it has a magnitude (30 m) and a direction (due north). Next, you walk 40 m due west. Once again, that is a vector. If you were to add the two vectors together, you would be asking the question, "If I restarted my journey but

---

[1] Remember that in quantum mechanics, an electron is not like the moon orbiting the earth but more like a wave that is surrounding the nucleus. We are just using the moon and earth as an analogy to introduce angular momentum.

[2] There are, at least, two things in physics that really boggle my mind. The first is spin. The second is something called the fine structure constant, which you should Google if you want your mind blown!

Fig. 7.1 An example of
adding two vectors together

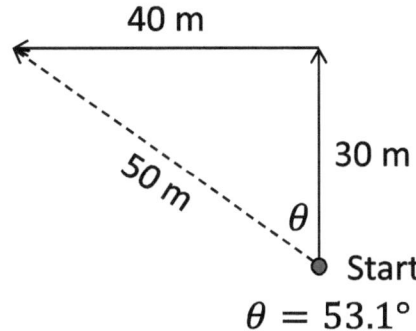

$$\theta = 53.1°$$

took the shortest path, how far and in what direction would I walk?" The answer to this particular scenario is 50 m at $\tan^{-1}\left(\frac{40\,\text{m}}{30\,\text{m}}\right) = 53.1°$ West of North.

> **The Super Important Take Home Message**
> Adding two vectors together produces a third vector. The size and direction of the third vector depends upon the sizes and directions of the first two vectors.

Let's get back to talking about angular momentum in a quantum system. An electron in an atom has a total angular momentum. That total angular momentum of the electron comes from the vector addition of its orbital angular momentum and its spin. There are quantum numbers associated with all three of these angular momentum vectors (orbital, spin, and total). The rest of this chapter is going to be focused on understanding angular momentum in a quantum system. We will find that both the size and direction of an angular momentum vector are quantized (i.e., can only have certain values like the size of an orbital angular momentum vector is $\sqrt{\ell(\ell+1)}\hbar$ where ℓ is a positive integer or zero). We will start by exploring the orbital angular momentum of a single electron in an atom and then build up the complexity by exploring (and adding) more and more angular momentum vectors to the system.

7.3   Orbital Angular Momentum of a Single Electron

The magnitude of the orbital angular momentum vector for a single electron in an atom is represented by the quantum number ℓ. The direction (or orientation) is also quantized and is represented by the quantum number m_ℓ. Figure 7.2a) shows an example of the three possible orientations of an orbital angular momentum vector represented by $\ell = 1$. Each of these vectors has the same magnitude (they all have $\ell = 1$), but they all point in different directions (they all have a different

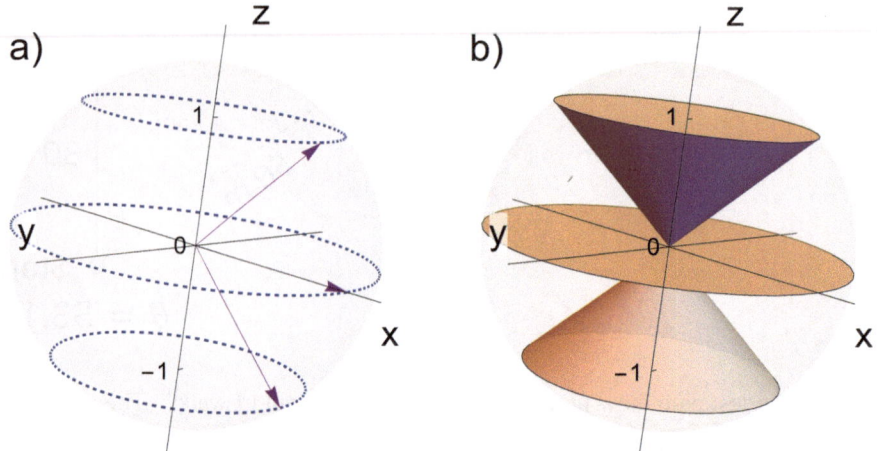

Fig. 7.2 Two popular ways to think about angular momentum states for a quantum mechanical system, like an electron in an atom. (**a**) A vector representation of the three possible vector orientations for an electron with orbital angular momentum quantum number $\ell = 1$. (**b**) The same information using cones instead of a vector and a dotted circle

m_ℓ quantum number). While ℓ tells us the magnitude of the vector, m_ℓ tells us how much of the vector points along the z-axis; this is called the z-component or z-projection. Figure 7.2a) has three values of m_ℓ: -1, 0, and $+1$. The dashed blue circles are meant to indicate that the angular momentum vector points anywhere on that circle. Because the angular momentum vector can point anywhere on the circle, some people like to think of angular momentum as a cone, see Fig. 7.2b).[3] These cones are also referred to as angular momentum states, which we will make more sense as we work our way through Chaps. 7 and 8.

Orientation of the angular momentum vector, height of the cone, or component/projection along the z-axis are all valid ways to think about m_ℓ. After solving the math for quantum mechanics, we find the component of orbital angular momentum along the z-axis to be $m_\ell \hbar$. For the upside-down cone in Fig. 7.2b), the orbital angular momentum vector is represented by the two quantum numbers $\ell = 1$ and $m_\ell = -1$, or the size of the orbital angular momentum is $\sqrt{1(1+1)}\hbar = \sqrt{2}\hbar$ and the amount of orbital angular momentum along the z-axis (or height of the cone) is $-\hbar$. The flat cone has the same size ($\ell = 1$), but it has no component along the z-axis ($m_\ell = 0$). Finally the upright cone has a component along the z-axis represented by $m_\ell = +1$. Since m_ℓ is the amount pointing along the z-axis, that number is also restricted by the value of ℓ. For example, if $\ell = 1$, m_ℓ cannot be 2. If it was, the amount pointing along the z-axis would be larger than the total magnitude!

---

[3] I, personally, like using cones. But, as long as you understand the concepts, it doesn't matter which version you use.

Experimentalists can measure these values. If an electron in an atom has orbital angular momentum represented by $\ell = 1$, the experimentalist will measure the size to be $\sqrt{2}\hbar$ every single time. However, if they measured the z-component, they could find $+\hbar$, 0, or $-\hbar$. But wait, couldn't they measure the x-component? What makes the z-component quantized while the x- and y- components just have to be somewhere along the dashed blue circle? On that note, how does an electron even decide which direction is z?

The general answer to all those questions is that we (the physicists) just decided to call one of the directions z. After we randomly pick a direction and call it z, we measure the amount along that axis. When we do that, we will only find those three values: $+\hbar$, 0, or $-\hbar$. We could certainly measure the x-component and get a value. In fact, if we measured the x-component we would get three possible values: $+\hbar$, 0, or $-\hbar$. The pictures in Fig. 7.2 would look very similar except the circles/cones would be aligned along the x-axis. Let's call those components $m_{\ell,x} = 1$, $m_{\ell,x} = 0$, and $m_{\ell,x} = -1$. The three possible values for the z-component form a basis set of size three. The three possible values for the x-component also form a basis set of size three, but those states making up that basis set are different than the state making up the basis set for the z-component. More importantly, those two basis sets are incompatible with each other.

Reminder of Important Definitions

- **Incompatible observables:** If two observables cannot be precisely measured at the same time, they are called incompatible. Measuring one observable changes the system, making subsequent measurements of the other observable unpredictable.
- **Compatible observables:** If two observables can be precisely measured at the same time, they are called compatible. Measuring one observable does not change the system in a way that affects the measurement of the other observable. Remeasuring the first observable will return the same value as initially measured.

Important

In the following paragraphs, all of the states (ψ) are for angular momentum. I am going to use ψ_z to represent the z-component of the angular momentum state, ψ_y for the y-component, and ψ_x for the x-component.

Suppose our system is in a state $\ell = 1$, $m_\ell = 1$. This is the upright cone in Fig. 7.2b). Since the x- and z- components are incompatible, we don't know what we will measure along the x-direction. We know there will be three possible outcomes for the amount of angular momentum along the x-axis ($+\hbar$, 0, or $-\hbar$), but we don't know which value we will find until we measure it. This concept is what is being represented by the blue circles or the cones.

Mathematically, we can also think about this using the principle of superposition. Each set of three cones forms a basis set. One basis set is the three cones representing orbital angular momentum along z and other is the three cones along x. A state from one basis set can be constructed from a superposition of the other basis set. For example,

$$\psi_z(\ell = 1, m_\ell = 1) = c_{x,1}\psi_{x,1} + c_{x,0}\psi_{x,0} + c_{x,-1}\psi_{x,-1}, \tag{7.1}$$

where $\psi_z(\ell = 1, m_\ell = 1)$ is the state representing that upright cone (orbital angular momentum along z), $\psi_{x,1}$, $\psi_{x,0}$, $\psi_{x,-1}$ are the three states making up the x-component of the orbital angular momentum basis set, and $c_{x,1}^2$, $c_{x,0}^2$, and $c_{x,-1}^2$ are the probabilities that, upon measurement of the x-component of angular momentum, the system has x-component values $m_{\ell,x} = 1$, $m_{\ell,x} = 0$, or $m_{\ell,x} = -1$ (remember the actual value of the x-component of orbital angular momentum is $m_{\ell,x}\hbar$). We could repeat the same argument with the y-component. In fact, all three components are incompatible with each other. To summarize this, here are the three basis sets:

$$\begin{aligned} x &: \psi_{x,1}, \psi_{x,0}, \psi_{x,-1} \\ y &: \psi_{y,1}, \psi_{y,0}, \psi_{y,-1} \\ z &: \psi_{z,1}, \psi_{z,0}, \psi_{z,-1} \end{aligned} \tag{7.2}$$

Each state represents a cone along a particular axis. We are certainly allowed to measure any component of angular momentum we want. Upon measurement, we will be in one of those states. However, since they are all incompatible observables, measuring one of the components puts the system in a superposition of the states for the other obervables. Interestingly, we have two separate ways to construct the state $\psi_z(\ell = 1, m_\ell = 1)$:

$$\begin{aligned} \psi_z(\ell = 1, m_\ell = 1) &= c_{x,1}\psi_{x,1} + c_{x,0}\psi_{x,0} + c_{x,-1}\psi_{x,-1} \\ \psi_z(\ell = 1, m_\ell = 1) &= c_{y,1}\psi_{y,1} + c_{y,0}\psi_{y,0} + c_{y,-1}\psi_{y,-1} \end{aligned} \tag{7.3}$$

In a way, this makes sense. If the quantum mechanical system is in the state $\psi_z(\ell = 1, m_\ell = 1)$ and we measure the x-component of angular momentum, we will get $m_{\ell,x} = 1$ with probability $c_{x,1}^2$, $m_{\ell,x} = 0$ with probability $c_{x,0}^2$, or $m_{\ell,x} = -1$ with probability $c_{x,-1}^2$. On the other hand, we could have decided to measure the y-component at which point we would need to construct the state from the y-basis set to predict what we would measure if we measured the y-component. I should point out that we can calculate those amplitudes using quantum mechanics. For this particular system, $c_{x,1} = 1/2$, $c_{x,0} = -1/\sqrt{2}$, and $c_{x,-1} = 1/2$.

Ok, so all three components are incompatible with each other, but what about compatibility with the magnitude of the angular momentum vector? Interestingly, the magnitude of the angular momentum vector is a compatible observable with all three components of angular momentum! At first, this might seem a little weird; how can all the components be incompatible with each other yet each component is compatible with the magnitude? But, in a way, this also makes physical sense. The length of the vector is the same whether the cones point along the x-direction, the y-direction, or the z-direction. Measuring a component of angular momentum does not change the length, but it does change the orientation. For this system, measuring the magnitude of angular momentum will always return $\sqrt{2}\hbar$ (or $\ell = 1$). It doesn't matter if we are in a state for the x-component, y-component, or z-component, $\ell = 1$.

The rest of the chapter will be spent making the system more complex and complete. To do this, we will need to pick one component of angular momentum to represent the system. Since the x-, y-, and z- components are just rotations of each other, it doesn't really matter which one we use to represent direction. By tradition, we pick the z-component.

In the end, the above discussion can be summarized as follows: for $\ell = 1$, there are three possible projections for an axis and they all look the same ($+\hbar$, 0, and $-\hbar$). So, we need to pick some axis to represent the idea that there are three orientations for the orbital angular momentum vector.

A lot has happened in this section, so let's summarize:

Summary

- Orbital angular momentum is quantized in two ways: the magnitude and the projection on an axis.
- The magnitude is represented by the quantum number ℓ, which is used to calculate the magnitude using the formula $\sqrt{\ell(\ell + 1)}\hbar$.
- The component of the orbital angular momentum along the z-axis is represented by the quantum number m_ℓ, which is used to calculate the component along the z-axis using the formula $m_\ell\hbar$.
- The choice of the z-axis is random, but we need an axis to represent the fact that there are only certain directions the vector can point along that axis.
- We can still measure along other directions, but the measurements along those directions are incompatible with measurements along the z-direction. We use cones to remind us that, while we know the z-component of angular moment, the x- and y-components are incompatible with the z-component.

7.4   The Magnitude and Projection of Angular Momentum

The magnitude of an orbital angular momentum vector is given by the formula $\sqrt{\ell(\ell+1)}\hbar$. The amount of the vector pointing along the z-axis (or the height of the cone) is $m_\ell\hbar$. Notice that the formulas contain only quantum numbers and the reduced Planck's constant. If you hear someone say, "Angular momentum comes in units of $\hbar$", this is what they mean. Being told that a quantum mechanical system has quantum numbers $\ell = 4$ and $m_\ell = -2$ tells us all we need to know about the orbital angular momentum of the system: the magnitude is $\sqrt{4(4+1)}\hbar = \sqrt{20}\hbar$ and the z-component is $-2\hbar$ (the cone is pointed in the negative z-direction with a height of $2\hbar$).

There are also mathematical restrictions on the quantum numbers themselves. ℓ can be either zero (no orbital angular momentum) or a positive integer (it has orbital angular momentum). The component along the z-axis must be smaller than the magnitude of the orbital angular momentum, so m_ℓ is restricted to be between $-\ell$ and $+\ell$ in integer steps.

Example #1 An electron in an atom has orbital angular momentum represented by the quantum number $\ell = 3$. There are 7 possible orientations (cone heights) of that vector represented by $m_\ell = -3, -2, -1, 0, 1, 2, 3$, see Fig. 7.3. If we measure the magnitude of the orbital angular momentum, we will measure $\sqrt{3(3+1)}\hbar = \sqrt{12}\hbar$. If we measure the z-components, we will measure either $-3\hbar, -2\hbar, -\hbar, 0, \hbar, 2\hbar$, or, $3\hbar$.

Example #2 An electron in an atom has orbital angular represented by $\ell = 2$. There are 5 possible orientations of that vector represented by $m_\ell = -2, -1, 0, 1,$ and 2.

Fig. 7.3 A vector representation of the seven possible vector orientations for an electron with orbital angular momentum quantum number $\ell = 3$

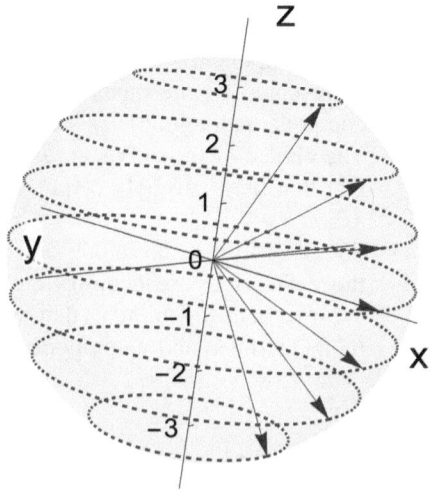

The form of $\sqrt{\ell(\ell+1)}\hbar$ and $m_\ell\hbar$ are the same for other types of angular momentum. All we have to do is replace the quantum numbers representing orbital angular momentum with the quantum numbers that represent the type of angular momentum we are interested in. In the next few sections, we are going to expand our discussion to talk about electron spin, the total electronic angular momentum (orbital + spin), the total orbital angular momentum of multiple electrons in an atom, and more. Each of the above examples of angular momentum will have two quantum numbers associated with it, one for magnitude and one for the z-component (cone height). For example, electron spin will have quantum numbers s and m_s. The formulas for electron spin are $\sqrt{s(s+1)}\hbar$ and $m_s\hbar$. While the quantum numbers for orbital angular momentum are integers, we will find that the other types of angular momentum quantum numbers can be either integers or half-integers. But they all use the same two formulas and tell us the same information, magnitude and z-component (cone height).

7.5 Adding Angular Momentum Vectors Together

Imagine you had 2 electrons in your atom. Each of those electrons has orbital angular momentum. We want to ask the question, "What is the *total* orbital angular momentum of the system?" To answer that question, we need to add together two quantum mechanical vectors. As an example, suppose we have an electron with the magnitude of the orbital angular momentum vector represented by $\ell_A = 1$ (Fig. 7.2 with 3 possible orientations) and a second electron with $\ell_B = 3$ (Fig. 7.3 with 7 possible orientations). We want to add these two vectors together as in Fig. 7.1, but we have to be thoughtful for two big reasons: (1) the vectors are really more like "vector cones" and (2) there might be incompatible observables we have to think about.

While the math is surprisingly complicated, the result is quite nice:

$$L = (\ell_A + \ell_B), (\ell_A + \ell_B - 1), (\ell_A + \ell_B - 2), \ldots, |\ell_A - \ell_B| \qquad (7.4)$$

where L is the quantum number for the composite system. For each value of L, there are $2L + 1$ possible projections represented by the quantum number m_L. This might be, understandably, confusing. So, let's unpack this statement using our example of $\ell_A = 1$ and $\ell_B = 3$.

If we add together two quantum mechanical vectors whose magnitudes are represented by quantum numbers ℓ_A and ℓ_B, the total angular momentum of the system will be represented by new quantum numbers L and m_L. The math shows there are multiple possible magnitudes one can get when adding two angular momentum vectors together. However, the new cones for L and m_L follow the same rules that we have been working with: L tells us the magnitude of the angular momentum for the system and m_L tells us the component (or height of the cone). From the math, the largest possible magnitude of L is given by $\ell_A + \ell_B$, which is $L = 4$ in the thought experiment we are working through. The smallest possible magnitude is $L = |\ell_A - \ell_B|$, or $L = 2$ for the example we are working on. To find all possible values of L, we make a list that starts with the smallest value and continually add 1 until we reach the largest value. In this case L can equal 2, 3, or 4.

To put it another way, there are three different orientations for ℓ_A and seven different orientations for ℓ_B. Depending on the orientations, adding those two angular momentum vectors results in different possible magnitudes. Working through all possible combinations results in three possible magnitudes. For each possible magnitude, there are different orientations (cone heights). In this example, the individual orbital angular momentum vectors can add together to make $L = 2$ (5 possible cones with the same length vector $\sqrt{2(2+1)}\hbar = \sqrt{6}\hbar$), $L = 3$ (7 possible cones with the same length vector $\sqrt{3(3+1)}\hbar = \sqrt{12}\hbar$), and $L = 4$ (9 possible cones with the same length vector $\sqrt{4(4+1)}\hbar = \sqrt{20}\hbar$).

After all the math, there are $5+7+9 = 21$ possible cones. This makes sense since the original vectors had 3 and 7 possible orientations, and $3 \times 7 = 21$. However, I want to caution you in that the new cones are not constructed simply by taking one cone chosen from $\ell_A = 1$ and one cone from $\ell_B = 3$ and adding them together. This is because the z-components of the individual electrons are incompatible with the z-component of the composite system. So, the new cones, which we will call the basis set for the composite system, are a superposition of all the cones from the individual electrons, which we call the individual electron basis set. The complicated math, which we skip in this book, tells us the amplitude or amount that we need from each of the individual electron basis set to construct the new composite system basis set. We will explore this more in Sect. 7.5.1.

Another example: Suppose $\ell_A = 3$ and $\ell_B = 5$. The largest value for L is $3 + 5 = 8$, and the smallest value is $|3 - 5| = 2$. Therefore, L can have values of 2, 3, 4, 5, 6, 7, or 8. For each value of L, there will be different orientations represented by the quantum number m_L. If $L = 3$, there are 7 possible vector cones represented by $m_L = 3, 2, 1, 0, -1, -2$, and -3.

Quick problem (the answers are in the footnotes[a]):
Suppose $\ell_A = 1$ and $\ell_B = 2$.

(a) How many projections are there for $\ell_A = 1$?
(b) How many projections are there for $\ell_B = 2$?
(c) How many possible cones can we add together from part (a) and part (b)? In other words, what is the size of the individual electron basis set?
(d) What values of L are possible?
(e) For each value of L, how many projections are there?
(f) Add up all the possible projections from part (e).

The answer to (c) should match the answer in (f).

[a](a) 3; (b) 5; (c) $3 \times 5 = 15$; (d) 1, 2, or 3; (e) 3, 5, or 7; (f) 15.

7.5.1   Compatible or Incompatible?

For just these two electrons, we have a bunch of quantum numbers floating around. Each quantum number corresponds to something we can measure. For these two electrons, we have six quantum numbers: ℓ_A, $m_{\ell,A}$, ℓ_B, $m_{\ell,B}$, L, and m_L. Which of these represent compatible observables? Is there something we can measure that will disrupt the value of another observable? As learners of quantum mechanics, this is important to keep track of.

All of the magnitudes are compatible with everything. If we know $\ell_A = 1$, $\ell_B = 3$, and $L = 2$, we can measure any of the observables represented by the six quantum numbers without disrupting future outcomes of magnitude measurements. However, the composite z-component m_L is incompatible with the z-components of the individual electrons, $m_{\ell,A}$ and $m_{\ell,B}$. Therefore, we can know either the z-component of the system or the z-components of the individual electrons, but not all three at the same time. The z-components of the two individual electrons are compatible with each other. So, if we know the z-component of the system and want to know what we might measure if we were to measure the z-component of the individual electrons, we have to write the state of the composite system as a superposition of states from the individual electron basis states. The concept remains the same; we would just have to do the math to find the amplitudes.

Generalization and Summary

The above discussion is true when adding *any* two quantum mechanical angular momentum vectors together. Suppose we have two angular momentum vectors **A** and **B**. Their magnitudes are represented by quantum numbers A and B, while their z-components are represented by m_A and m_B. We want to add **A** and **B** together to get **C**. The possible values for the quantum number representing the magnitude of **C** are:

$$C = (A + B), (A + B - 1), \ldots, |A - B| \qquad (7.5)$$

For each value of C, $m_C = -C, \ldots, C$ in integer steps. The compatibility of the observables is as follows:

- The magnitudes are always compatible with everything.
- The z-components of the individual angular momenta are compatible with each other.
- The z-component of the summed angular momentum is incompatible with the z-components of the individual angular momenta.

7.6   Other Types of Angular Momentum

An electron, whether it is in an atom or free, always has spin. The quantum numbers for spin are s and m_s for the magnitude and orientation (cone height), respectively. For an electron, $s = 1/2$ always, so the possible orientations are always $m_s = -1/2$ or $m_s = +1/2$. From tradition, we call $m_s = +1/2$ "spin up" and $m_s = -1/2$ "spin down".

If the electron is in an atom, it can also have orbital angular momentum. We can add orbital angular momentum and spin together and ask the question, "What is the total angular momentum of that electron?" This new total angular momentum of the electron is represented by the quantum numbers j and m_j. Using our new rule (Eq. 7.5), the possible magnitudes of the total are represented by $j = l+s, \ldots, |l-s|$ in integer steps. Thus, if $\ell = 1$ and $s = 1/2$, j can be 3/2 or 1/2. As before, m_ℓ can be any value between $-\ell$ and $+\ell$ in integer steps, m_s can be any value between $-s$ and $+s$ in integer steps, and m_j can be any value between $-j$ and $+j$ in integer steps. Knowing the quantum numbers j, ℓ, and s tells us the magnitude of the angular momentum vectors. Knowing m_j, m_ℓ, and m_s tells us the orientation of each of their respective angular momentum vectors. However, m_j is incompatible with m_s and m_ℓ.

Here is a table summarizing all of the current types of angular momenta for a single electron in an atom (QN means quantum number):

Type	QN	Rule	Formula
Orbital	ℓ	Zero or positive integer	Magnitude: $\sqrt{\ell(\ell+1)}\hbar$
	m_ℓ	$-\ell$ to $+\ell$ in integer steps	Cone height: $m_\ell \hbar$
Spin	s	For a single electron, $s = 1/2$	Magnitude: $\sqrt{s(s+1)}\hbar = \sqrt{3/4}\hbar$
	m_s	$-1/2$ and $+1/2$	Cone height: $m_s \hbar$
Total electronic	j	Zero, positive integer, or half-integer	Magnitude: $\sqrt{j(j+1)}\hbar$
	m_j	$-j$ to $+j$ in integer steps	Cone height: $m_j \hbar$

Example An electron in an atom has quantum numbers $\ell = 3$ and $s = 1/2$.

(a) What is the magnitude of the electronic orbital angular momentum?
   Solution: The magnitude is $\sqrt{\ell(\ell+1)}\hbar = \sqrt{12}\hbar \approx 3.46\hbar$
(b) What is the magnitude of the electron's spin?
   Solution: The magnitude is $\sqrt{s(s+1)}\hbar = \sqrt{\frac{3}{4}}\hbar \approx 0.87\hbar$
(c) What are the possible magnitudes of the total electronic angular momentum?
   Solution: The possible values for j are $j = (\ell+s), \ldots, |\ell-s|$ in integer steps. The largest value is $3 + \frac{1}{2} = \frac{7}{2}$ and the smallest value is $|3 - \frac{1}{2}| = \frac{5}{2}$. So, $j = \frac{7}{2}$ or $j = \frac{5}{2}$.

When $j = 7/2$, the magnitude is $\sqrt{j(j+1)}\hbar = \sqrt{\frac{7}{2}\frac{9}{2}}\hbar = \sqrt{\frac{63}{4}}\hbar \approx 3.97\hbar$.

When $j = 5/2$, the magnitude is $\sqrt{j(j+1)}\hbar = \sqrt{\frac{5}{2}\frac{7}{2}}\hbar = \sqrt{\frac{35}{4}}\hbar \approx 2.96\hbar$.

We now have all the building blocks we need to add together as many angular momentum vectors as we need. We first add two together to come up with the new possible quantum numbers and projections for this composite system. Next, we treat the composite system as its own individual angular momentum and add a third angular momentum. Then keep going until you have accounted for all electrons in your atom. The magnitudes of each angular momenta will be compatible with everything. The z-component of any summed angular momentum will be incompatible with the two individual angular momenta.

In practice, we don't go through all this effort. As we will see in Chap. 8, there are shortcuts we can take and databases that tell us all we need to know. For atoms with more than 1 electron, we do not keep track of subscripts (i.e., electron 1, electron 2, etc.), we simply ask the question "how much orbital, spin, and total electronic angular momentum do all of the electrons have?" The formulas and ideas are identical, but we use capital letters when we ask how much angular momentum the electrons as a whole have:

Type	QN	Rule	Formula
Orbital	L	Zero or positive integer	Magnitude: $\sqrt{L(L+1)}\hbar$
	m_L	$-L$ to $+L$ in integer steps	Cone height: $m_L\hbar$
Spin	S	Zero, positive integer, or half-integer	Magnitude: $\sqrt{S(S+1)}\hbar$
	m_S	$-S$ to $+S$ in integer steps	Cone height: $m_S\hbar$
Total electronic	J	Zero, positive integer, or half-integer	Magnitude: $\sqrt{J(J+1)}\hbar$
	m_J	$-J$ to $+J$ in integer steps	Cone height: $m_J\hbar$

All of the magnitudes are compatible with everything. However, m_J is incompatible with m_L and m_S.

Something Super Annoying, But Unfortunately Super Important Historically, we assign letters to orbital angular momentum quantum numbers. A system with no orbital angular momentum has the quantum number $L = 0$ (or $\ell = 0$ for a single electron) is given the letter designation S (or s for a single electron). But, S and s are also the quantum numbers for spin. They are both angular momenta, but very different types. Technically the quantum numbers are italicized and the letter designations are not, but that is not a rule everyone follows. Below is a table with quantum numbers, the letter designation, and double uses with other types of angular momentum:

Orbital quantum number	Letter designation	Other uses
$L = 0$ ($\ell = 0$)	S (s)	S and s are also used for spin
$L = 1$ ($\ell = 1$)	P (p)	
$L = 2$ ($\ell = 2$)	D (d)	
$L = 3$ ($\ell = 3$)	F (f)	F is also used for nuclear angular momentum
$L = 4$ ($\ell = 4$)	G (g)	

I admit it is extremely annoying to have orbital angular momentum represented by letters that also mean different types of angular momentum, but it is unfortunately the way of spectroscopy. It takes some getting used to.

The Nucleus The nucleus can also have angular momentum. Traditionally, this is called nuclear spin, but that is a bit of a misnomer. The angular momentum of the nucleus comes from the spin of the protons and neutrons as well as any orbital angular momentum of those nucleons. So, nuclear spin should really be called total nuclear angular momentum but no one uses that phrase. Nuclear spin has the quantum numbers I and m_I. Nuclear spin (I) can be added to the total electronic angular momentum (J) find the total angular momentum of the atom. The total angular momentum is represented by the quantum numbers F and m_F, where F is found from J and I using Eq. 7.5:

$$F = (J + I), (J + I - 1), \ldots, |J - I| \qquad (7.6)$$

As always, m_F is incompatible with m_J and m_I.

So, if an atom has nuclear spin, our table gets a bit longer:

Type	QN	Rule	Formula
Orbital	L	Zero or positive integer	Magnitude: $\sqrt{L(L+1)}\hbar$
	m_L	$-L$ to $+L$ in integer steps	Cone height: $m_L\hbar$
Spin	S	Zero, positive integer or half-integer	Magnitude: $\sqrt{S(S+1)}\hbar$
	m_S	$-S$ to $+S$ in integer steps	Cone height: $m_S\hbar$
Total electronic	J	Zero, positive integer, or half-integer	Magnitude: $\sqrt{J(J+1)}\hbar$
	m_J	$-J$ to $+J$ in integer steps	Cone height: $m_J\hbar$
Nuclear spin	I	Zero, positive integer, or half-integer	Magnitude: $\sqrt{I(I+1)}\hbar$
	m_I	$-I$ to $+I$ in integer steps	Cone height: $m_I\hbar$
Total Atomic	F	Zero, positive integer, or half-integer	Magnitude: $\sqrt{F(F+1)}\hbar$
	m_F	$-F$ to $+F$ in integer steps	Cone height: $m_F\hbar$

We will explore nuclear spin more in Chap. 9.

The Photon The photon has intrinsic angular momentum as well. Photon spin is the quantum mechanical description of light polarization, and it has a quantum number of 1. In Chap. 1, we talked about how light can be linearly polarized, circularly polarized, or elliptically polarized. That statement is for a laser beam, which is composed of many photons. A single photon is circularly polarized. To help visualize this, imagine a photon traveling straight towards you. The photon would be rotating either clockwise or counterclockwise. To be clear, a photon is not actually rotating just like the electron is not actually spinning like a top. However, both the electron and the photon behave and interact with other particles as if they are spinning or rotating. If it is spinning counterclockwise, the height of the cone is $+\hbar$ (the z-component to photon spin with a quantum number of +1). If the photon is spinning clockwise, the height of the cone is $-\hbar$.[4]

Circularly polarized light just means that all the photons are rotating in the same direction. A laser beam composed of photons coming towards you that are all rotating clockwise is said to have left (or left-handed) circularly polarized light. A laser beam composed of photons that are all rotating counterclockwise is said to have right (or right-handed) circularly polarized light. To get linearly polarized light, you need to have equal amounts of photons spinning clockwise and counterclockwise. Elliptically polarized light has an imbalance between the number of clockwise and counterclockwise photons.

7.7   A Bit More on Compatible and Incompatible Observables

Every quantum number in every table in Sect. 7.6 is something we can measure. Consider the hydrogen atom, which has a single electron. For now, we will ignore the fact that the nucleus of a hydrogen atom has angular momentum (this is the topic of Chap. 9 so we will revisit compatible and incompatible observables again in that chapter). Here is a list of things we can measure: ℓ, m_ℓ, s, m_s, j, and m_j.

Technically, we could also measure L, m_L, S, m_S, J, and m_J, but, since the hydrogen atom has only a single electron, measuring, for example, the total orbital angular momentum of all the electrons is the same as measuring the orbital angular momentum of the single electron in the system.

We want to ask the question, which of these observables is compatible with energy? In other words, if the hydrogen atom was in an energy state, what other properties could we measure and still leave the electron in that same energy state? The answer is:

- Compatible with energy: ℓ, s, j, and m_j
- Incompatible with energy: m_ℓ and m_s

That means if the system was in an energy state and we measured the z-component of the electron's orbital angular momentum, the system is now in a superposition

---

[4] Because the photon has no mass, the math, interestingly, forbids a z-component of 0.

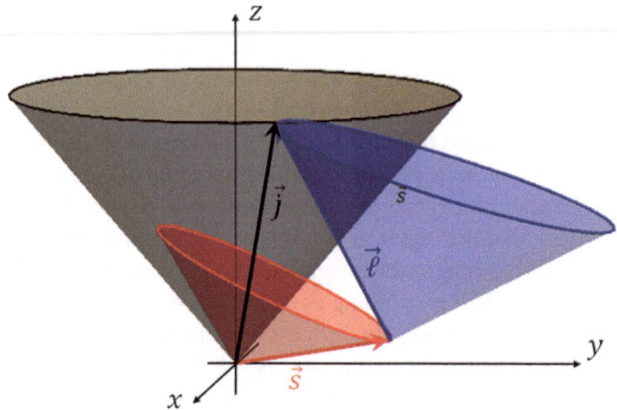

Fig. 7.4 An example of adding together a spin cone and a orbital angular momentum cone to get the total electronic angular momentum cone. Notice that the z-component of both the spin and orbital angular momentum cones is not the same value at all points of the cone tops. This means the z-component does not have a specific, discrete value, which means that both the z-component of spin and the z-component of orbital angular momentum are incompatible with the z-component of j

of energy states. But what makes m_ℓ and m_s incompatible with m_j and energy? While it isn't obvious why m_ℓ and m_s are incompatible with the energy state, we can visualize why they are incompatible with m_j. Figure 7.4 visualizes the incompatibility. This figure shows the addition of two angular momentum cones. Since m_j is compatible with the energy state, we will define the z-axis to be along the cone for j. I want to point out a few things. First, notice that the cones for both ℓ and s are tilted with respect to the z-axis. This means that the height of those two cones (i.e., the result of a measurement of m_ℓ or m_s) are no longer perfect projections onto the z-axis. In other words, if we measured the z-component of spin with respect to this z-axis, we will be in a superposition of the m_j basis set. Compare this to the measurement of the z-component of j. The cone for j has a rim that is constant along the z-axis, so we will always measure the same value of m_j. Since m_j is compatible with the energy state, measuring m_ℓ or m_s would also put the system into a superposition of energy states.

The second thing I would like to point out is that the length of the both s and ℓ are constants. If we were to measure the slant height of either cone, we will always get the same answer. In other words, both s and ℓ are compatible with j and energy.

To clarify which observables are compatible with the energy states, I'll reintroduce the quantum number for energy, n, from Chap. 6. We use the quantum numbers representing those observables that are compatible with energy in a ket: $|n\, \ell\, s\, j\, m_j\rangle$. We will discuss these quantum numbers and their relationships to electron shells in more detail in Chap. 8. If the electron in the hydrogen atom is in an energy state represented by one of these kets, we can measure any of those observables and still be in that same energy state. If we measured m_ℓ or m_s, the system would be in a superposition of energy states.

Let's solidify this idea with a couple of examples. We will do Example #1 together, and then you should do Example #2.

Example #1 How many states will there be for the electron in a hydrogen atom that has $n = 2$, $\ell = 1$, and $s = 1/2$?

To start this problem, let's first calculate what values of j are possible. j can range from $\ell + s$ to $|\ell - s|$ in integer steps, so j can be 1/2 or 3/2. If $j = 1/2$, then m_j can be 1/2 or $-1/2$. If $j = 3/2$, then m_j can be 3/2, 1/2, -1/2, or -3/2. Therefore, there are 6 energy states with $n = 2$, $\ell = 1$, and $s = 1/2$. Using the notation $|n\,\ell\,s\,j\,m_j\rangle$, these states are: $|2\,1\,\frac{1}{2}\,\frac{1}{2}\,\frac{1}{2}\rangle$, $|2\,1\,\frac{1}{2}\,\frac{1}{2}\,-\frac{1}{2}\rangle$, $|2\,1\,\frac{1}{2}\,\frac{3}{2}\,\frac{3}{2}\rangle$, $|2\,1\,\frac{1}{2}\,\frac{3}{2}\,\frac{1}{2}\rangle$, $|2\,1\,\frac{1}{2}\,\frac{3}{2}\,-\frac{1}{2}\rangle$, and $|2\,1\,\frac{1}{2}\,\frac{3}{2}\,-\frac{3}{2}\rangle$.

The above math tells us that there is an energy state in the hydrogen atom that is represented by, for example, the quantum numbers $|2\,1\,\frac{1}{2}\,\frac{3}{2}\,-\frac{3}{2}\rangle$. If the hydrogen atom is in this state, it has a defined magnitude of orbital angular momentum (in this case $\sqrt{\ell(\ell+1)}\hbar = \sqrt{1(1+1)}\hbar \approx 1.41\hbar$), a defined magnitude for spin, a defined magnitude for the total electronic angular momentum, and a defined z-component for the total electronic angular momentum. That state does not have a well defined z-component of orbital angular momentum nor a well defined z-component of spin. We can measure any observable compatible with energy and the system will stay in the $|2\,1\,\frac{1}{2}\,\frac{3}{2}\,-\frac{3}{2}\rangle$ state. However, if we measure m_ℓ or m_s, the system will be in a superposition of energy states.

Example #2 The electron in the hydrogen is in the state $|2\,1\,\frac{1}{2}\,\frac{3}{2}\,-\frac{1}{2}\rangle$.

(a) What **will** we measure for the electron's orbital angular momentum, the electron's spin, the electron's total angular momentum, and the electron's z-component of the electron's total angular momentum?
(b) If we measured the z-component of the electron's orbital angular momentum, what might we get?

The solutions are below the crossword.

Across:

2 ____ - ____ notation is a common way to show all the quantum numbers that are compatible with energy.

6 Applying a force changes this

7 ℓ is an example of a quantum ____

Down:

1 Apply this to change an object's angular momentum.

3 Apply this to change an object's momentum.

4 The cone with height $+\hbar$ in Figure 7.2 is an example of an angular momentum ____

5 Applying a torque changes an object's ____ momentum.

Solution

(a) $\ell = 1$: We will measure $\sqrt{1(1+1)}\hbar \approx 1.41\hbar$ with 100% probability

   $s = 1/2$: We will measure $\sqrt{\frac{1}{2}(\frac{1}{2}+1)}\hbar \approx 0.87\hbar$ with 100% probability

   $j = 3/2$: We will measure $\sqrt{\frac{3}{2}(\frac{3}{2}+1)}\hbar \approx 1.94\hbar$ with 100% probability

   $m_j = -1/2$: We will measure $-\frac{1}{2}\hbar$ with 100% probability

(b) We would measure either $-\hbar$, 0, or $\hbar$ with different probabilities. We would need advanced math to calculate those probabilities, so I will just emphasize that we will not measure a definite value for the z-component of the electron's orbital angular momentum. After the measurement, the system will no longer be in the state $|2\ 1\ \frac{1}{2}\ \frac{3}{2}\ \text{-}\frac{1}{2}\rangle$, but a superposition of energy states. If we now went back and measured j, we might get $j = 1/2$ or $j = 3/2$ with hard to calculate probabilities.

Problems

7.1 Angular momentum seems to be a pretty important concept in quantum mechanics and atomic physics! In your own words, describe a classical (i.e., not quantum) system that has angular momentum and a classical system that does not. Speculate on a few ways the classical system with angular momentum might behave differently if it were a quantum system.

7.2 An electron in an atom has the quantum numbers $\ell = 0$ and $s = 1/2$.

(a) What are the magnitudes and projection along the z-axis for ℓ?
(b) What are the magnitudes and projection along the z-axis for s?
(c) What are the possible magnitudes and projections along the z-axis for j?

7.3 An atom has two electrons. One electron has $\ell = 1$ and the other has $\ell = 0$.

(a) Find all possible values of L.
(b) Find all possible values of S.
(c) Find all possible values of J.

Hint: There will be a different set of J values for each combination of L and S. For example, if your answer to part (a) was $L = 2$ or $L = 1$ (I hope you didn't get these numbers, because they aren't correct) and your answer to part (b) was $S = 0$ or $S = 1$, then there would be 4 sets of answers to part (c). Word your answer similar to:

"For $L = 2$ and $S = 0$, J can be ..."
"For $L = 2$ and $S = 1$, J can be ..."

"For $L = 1$ and $S = 0$, J can be …"
"For $L = 1$ and $S = 1$, J can be …"

7.4 An atom has two electrons. One electron has $\ell = 1$ and the other has $\ell = 2$.

(a) Find all possible values of L.
(b) Find all possible values of S.
(c) Find all possible values of J.

7.5 The atom in Problem 3 has a nuclear spin of $I = 3$. Find all possible values of F.
Hint: There will be a different set of F values for each J value.

Electronic Structure and Atomic Notation

8

Abstract

In this chapter, we explore the fundamental principles governing the electronic structure of atoms and the notation used to describe it. We begin by examining energy level spacings across various elements, highlighting how the number of electrons influences the complexity of these levels. The Coulomb interaction and electron shell filling patterns are introduced as key factors determining atomic state energies. Through detailed discussions on electronic configurations and term symbols, we illustrate how angular momentum, both in individual electrons and in atoms as a whole, impacts energy states. We also discuss fine structure splitting and hyperfine structure splitting. By the end of this chapter, readers will have a comprehensive understanding of the labels used by atomic physicists for electronic levels in atoms. Additionally, we differentiate between fermions and bosons, emphasizing their roles and significance in atomic and nuclear physics.

Learning Goals

By the end of this chapter, you should be able to understand:

- the Coulomb interaction and its role in determining the energy levels within an atom.
- describe how electrons fill shells and subshells according to quantum mechanical principles.
- the concept of angular momentum in the context of atomic physics and how it affects atomic energy states.

© The Author(s) 2025
W. Raven, *Atomic Physics for Everyone*,
https://doi.org/10.1007/978-3-031-69507-0_8

- the influence of Coulomb interactions, electron shell filling, and angular momentum on the energy of atomic states.
- interpret and describe electronic configurations and term symbols to understand the angular momentum of individual electrons and the overall atom.
- explain the difference between fine structure splitting and hyperfine structure splitting.
- distinguish between fermions and bosons and explain their significance in atomic and nuclear physics.

8.1   Energy Level Spacings

Figure 8.1 shows the energy levels for hydrogen, helium, lithium, and europium. As a reminder, hydrogen has 1 electron, helium has 2, lithium has 3, and europium has 63. Notice how different the energy levels are! Hydrogen's first excited state is over $80,000\,\mathrm{cm}^{-1}$ above the ground state, while helium's first excited state is around

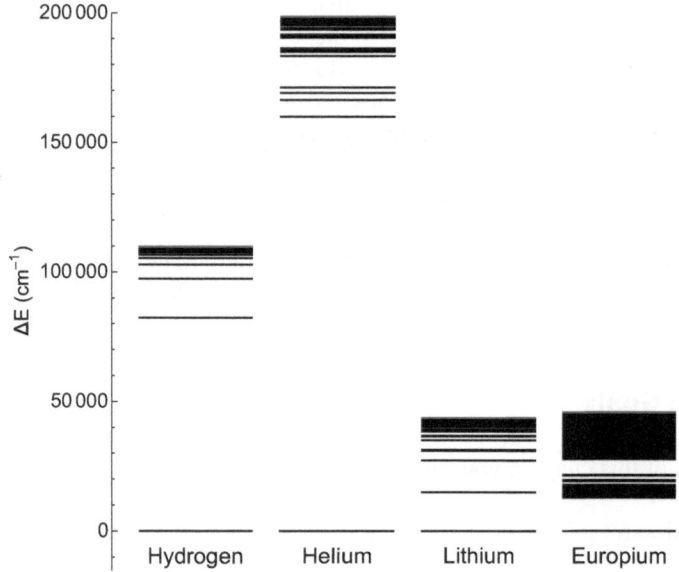

Fig. 8.1 The energy levels of hydrogen, helium, lithium, and europium. The red line at the top of each element is the ionization threshold, which is the energy required to remove an electron from the atom

Fig. 8.2 A zoom in on some
of the denser states in
Europium. Even though there
are many states, they are still
discrete

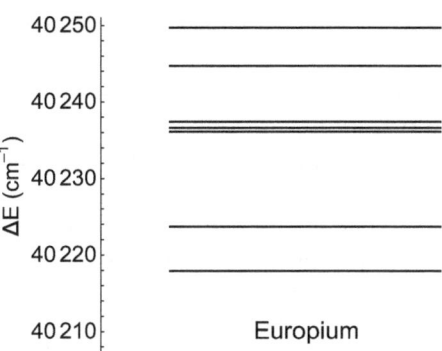

$160,000\,\text{cm}^{-1}$ above the ground state![1] The hydrogen states seem to get closer and closer to each other as the energy increases, but helium seems to "clump" a bit more. And look at Europium; there seems to be a big gap between the ground state before a really dense set of energy levels, a gap, and then even more! The red line on each element is the energy needed to rip an electron from the atom. We call this the **ionization threshold**.

Even though the energy levels are so densely packed, they are still discrete. The energy levels looking like a solid band is just an artifact of making a picture with many, many discrete energy levels (the europium diagram has 500 levels in it). Figure 8.2 is a zoom in of the Europium energy levels from $40,210\,\text{cm}^{-1}$ to $40,250\,\text{cm}^{-1}$, which is right in the middle of one of the dense patches. As you can see, these 7 levels are very close together, but still discrete. Plotting all 500 levels together really highlights the groupings.

Guiding Question
Why are the energy levels so different for each element?

Give it some thought and then read on.

---

[1] Remember that the ground states of hydrogen, helium, lithium, and europium do not have 0 energy. Each element's ground state has an energy like the lowest mode of a standing wave has energy. The actual ground state energy of hydrogen is very different than that of helium or lithium. But, we always do spectroscopy with respect to the element's ground state energy. Figure 8.1 is a nice learning tool, but it can be misleading by giving the impression that, for example, the first excited state of lithium has a similar energy to the first excited state of europium. They do not!! The energy difference compared to their ground state is similar. A subtle, but important difference. ☺

8.2   The Coulomb Interaction and Electron Shells

Definitions
- **Coulomb force:** The force between two charged particles, described by $F = k_e \frac{q_1 q_2}{r^2}$, where F is the force, $k_e = 8.987 \times 10^9$ N $\cdot$ m^2/C^2 is Coulomb's constant, q_1 and q_2 are the charges, and r is the distance between them. It represents the attraction or repulsion between the charged particles.
- **Coulomb interaction:** The interaction between two charged particles due to their electric charges. It is governed by the Coulomb force and is fundamental in understanding the behavior of charged particles.

We use the phrase Coulomb interaction when we are thinking about general interactions between charged particles, such as how these interactions can affect energy levels. We use the phrase Coulomb force when we are specifically thinking about the force one charged particle exerts on the other.

There are many interactions that happen inside the atom. Each of these interactions affects where the discrete energy levels end up. If we, as physicists, understand all of these interactions we should be able to calculate properties of the atom like the energy of a particular state, the natural linewidth of each transition, and what would happen if we put the atoms in a particular situation such as inside an electric field, a magnetic field, or if we put a bunch of these atoms together so strongly compressed under gravity that the gravitational energy is converted into heat and eventually the atoms fuse together.[2] While there are many interactions happening in the atom, the two biggest contributions that determine the energy of an atomic state are the Coulomb interaction and how electrons fill up "shells."

The **Coulomb interaction** is how we describe the interaction between charged particles. For the hydrogen atom, there is only 1 Coulomb interaction. That interaction is between the single electron, which has negative charge, and the single proton, which has positive charge. In fact, this interaction is so simple we can exactly solve the Schrödinger equation to find the effect this interaction has on the energy levels. The helium atom has two electrons and two protons. The two protons are inside the nucleus and, while there is technically a Coulomb interaction between the two protons, that interaction is overwhelmed by the strong nuclear force, which is a topic in Chap. 10. Therefore, we can think of the nucleus as a single particle with a charge of $+2$. The two electrons on the other hand are not confined tightly together like the two protons in the nucleus. For the helium atom, there are effectively 3 Coulomb interactions: one between the two electrons, one between one of the electrons and the nucleus, and one between the other electron and the nucleus. The increased number of interactions causes most of the dramatic difference between the energy levels of hydrogen and helium.

---

[2] That was a really complicated way of saying "a star."

Fig. 8.3 How electrons fill shells. This diagram is often called Madelung energy ordering rule, named after the German physicist Erwin Madelung. The top row is the $n = 1$ shell and has one subshell. The second row is the $n = 2$ shell and has two subshells. The third row is the $n = 3$ shell and has three subshells, and so on

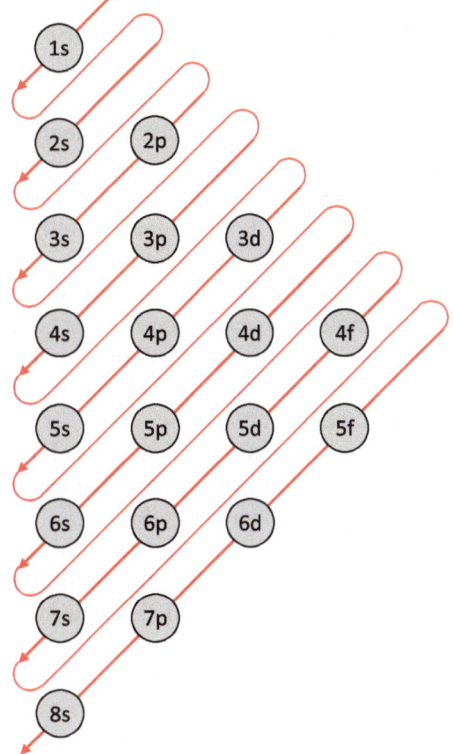

Lithium has 3 electrons and europium has 63. Like helium, we can think about the nucleus of lithium as 1 particle with a charge of +3. Even though europium has 63 protons in the nucleus, we can still think of the nucleus as 1 particle with a charge of +63.[3] Now we have a lot of interactions. Each electron has a Coulomb interaction with every other electron and each electron has a Coulomb interaction with the nucleus. One of the homework problems for this chapter will have you count the number of Coulomb interactions for the first few atoms in the periodic table.

The second major contribution comes from something called **electron shells**, which are made up of **electron subshells**. The first shell is called 1, and it contains 1 subshell labelled 1s, see Fig. 8.3. The second shell is 2 and has two shells. The third shell is 3 and has 3 subshells and so on. In Fig. 8.3, we stop showing all possible subshells on shell 5. There is a 5g subshell, 6f, 6g, 6h, etc. that is not shown. In quantum mechanics and atomic physics, the shell number (1, 2, 3, ...)

---

[3] The neutrons in the nucleus also matter, but not as much as the proton charge. The effect the neutrons have on the energy states is a topic in Chap. 10.

is a quantum number called the principal quantum number, represented by the letter n. This is the same quantum number n explored in Chap. 6.

The electron subshells come about because an electron in an atom can have orbital angular momentum and spin. The letter accompanying the principal quantum number is the orbital angular momentum quantum number for the electron in that subshell. The subshells use the historical letter designations discussed in Chap. 7. As a reminder, if an electron has the label s, it has $\ell = 0$. If it has a label p, it has $\ell = 1$, and so on. For example, 3p tells us that $n = 3$ and $\ell = 1$. From solving the Schrödinger equation, we find that n is the upper limit on possible values for ℓ. As a reminder, ℓ is a zero or a positive integer. From the math, we find $\ell_{max} = n - 1$. For example, if $n = 3$, then $\ell = 0$, 1, or 2. That means the $n = 3$ shell has three subshells: 3s, 3p, and 3d. There is no 3f subshell since $\ell = n$, which is not allowed.

An s-subshell can hold 2 electrons in total, one will have spin up ($m_s = +1/2$) and the other spin down ($m_s = -1/2$). A p-subshell can hold 6 electrons in total: 3 spin up and 3 spin down. A d-subshell can hold 10 electrons (5 up and 5 down), an f-subshell can hold 14, a g-subshell can hold 18, etc. The subshells fill in a specific order roughly shown in Fig. 8.3.

The first subshell to fill is the 1s subshell, which can hold 2 electrons. After the 1s subshell is filled, electrons start to fill the 2s subshell, which also holds 2 electrons. Once the 2s subshell is filled, electrons start to fill the 2p subshell, which holds 6 electrons. Next is 3s followed by 3p, 4s, 3d, 4p, etc.

Why Can a p-Shell Hold 6 Electrons? An electron with the label p means that $\ell = 1$. There are three possible orientations for this angular momentum vector: $m_\ell = -1$, 0, and 1. In addition, the electron can be either spin up or spin down. That means there are 6 orientations for an electron to have $\ell = 1$. Using the notation (m_ℓ, m_s), the possible orientations for a p-subshell are:

$$\left(1, +\tfrac{1}{2}\right) \quad \left(1, -\tfrac{1}{2}\right) \quad \left(0, +\tfrac{1}{2}\right) \quad \left(0, -\tfrac{1}{2}\right) \quad \left(-1, +\tfrac{1}{2}\right) \quad \left(-1, -\tfrac{1}{2}\right)$$

An electron with the label d means that $\ell = 2$. There are 5 possible orientations for this vector ($m_\ell = $ -2,-1,0,1,and 2). Including spin up and spin down for each m_ℓ gives 10 possible orientations. Using the notation (m_ℓ, m_s), the possible orientations for a d-subshell are:

$$\begin{pmatrix} 2, +\tfrac{1}{2} \\ 2, -\tfrac{1}{2} \end{pmatrix} \quad \begin{pmatrix} 1, +\tfrac{1}{2} \\ 1, -\tfrac{1}{2} \end{pmatrix} \quad \begin{pmatrix} 0, +\tfrac{1}{2} \\ 0, -\tfrac{1}{2} \end{pmatrix} \quad \begin{pmatrix} -1, +\tfrac{1}{2} \\ -1, -\tfrac{1}{2} \end{pmatrix} \quad \begin{pmatrix} -2, +\tfrac{1}{2} \\ -2, -\tfrac{1}{2} \end{pmatrix}$$

Figure 8.4 are Bohr model pictures of hydrogen, helium, and lithium in their lowest energy state (the ground state). Remember that electrons are actually more like waves, but I like to draw them as little balls to explore this concept. Hydrogen has 1 electron so the 1s subshell is half filled. Helium has 2 electrons that fully fill

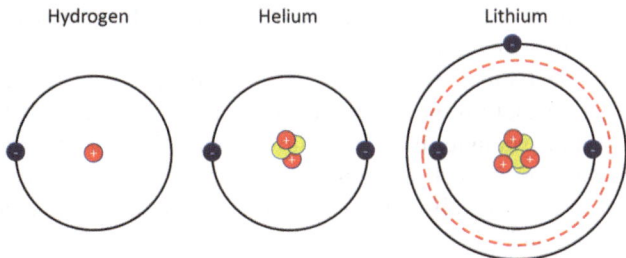

Fig. 8.4 A Bohr model picture of hydrogen, helium, and lithium. Electrons will first fill the 1s subshell. Hydrogen half fills the 1s subshell, while Helium fills the 1s subshell completely. Lithium has 3 electrons that completely fill the 1s subshell leaving a single electron in the 2s subshell

this 1s subshell. If we move up the periodic table to lithium (3 electrons), 2 electrons fill the 1s shell while the last electron is in the 2s shell.

So, why does this affect the energy levels in the atom? The answer is every time we fill a shell, it is like creating a new nucleus with a reduced charge. For example, let's look at a picture of lithium in Fig. 8.4. I added a red dashed line between the 1s and 2s shells. Inside this red dashed line, the effective charge is +1 (2 electrons and 3 protons). For all it knows, that 1 electron sitting in the 2s subshell is all by itself interacting with a nucleus with charge +1. This is just like the hydrogen atom with a few key differences. The first is the nucleus of lithium has a lot more mass than the nucleus of hydrogen. The second is that the outermost electron in lithium is farther away from the nucleus. The reduced Coulomb force means the energy levels for lithium are not as widely spaced compared to hydrogen. This is why the hydrogen atom and the lithium atom energy levels look so similar, but the lithium energy levels are closer together, see Fig. 8.5. All of the atoms in the first column of the periodic table (hydrogen, lithium, potassium, rubidium, cesium, and francium)

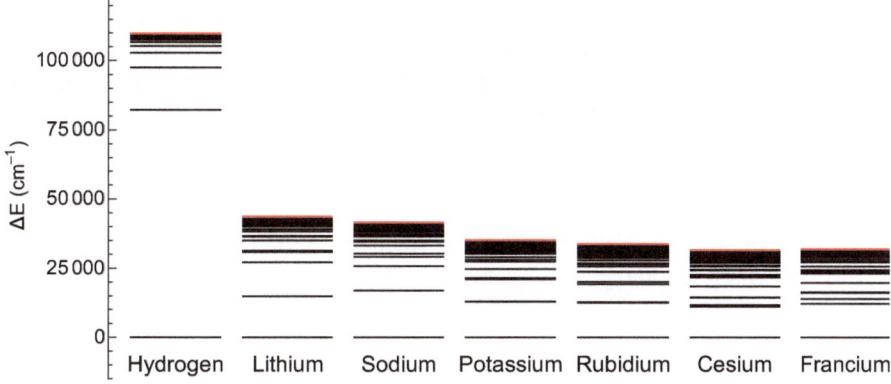

Fig. 8.5 The energy levels for the first column of the periodic table. Each of these elements has fully closed shells except for a single electron in an s-shell

have 1 electron sitting outside a closed shell. Therefore, the energy levels of these atoms are very similar to hydrogen.

We write the **electron configuration** with superscripts that tell us how many electrons are in a subshell; if there is no superscript present, there is only 1 electron in that subshell. For example, the electronic configuration for the ground state of helium is $1s^2$; there are 2 electrons in the 1s subshell. If we excite the atom so that 1 electron moves to the 3d subshell, the electronic configuration is 1s3d; 1 electron is in the 1s subshell and 1 is in the 3d subshell.

As we move through the periodic table, the atoms gain more and more electrons filling up shells according to Fig. 8.3. I should note that while the chart does a really good job, particularly for lighter atoms, it messes up sometimes for heavier atoms. For example, the chart predicts that the ground state of silver (47 electrons) should have a filled 5s subshell and 9 electrons in the 4d subshell:

$$1s^2 2s^2 2p^6 3s^2 3p^6 3d^{10} 4s^2 4p^6 5s^2 4d^9$$

However, the actual electronic configuration has a completely filled 4d subshell and a single electron in the 5s subshell:

$$1s^2 2s^2 2p^6 3s^2 3p^6 3d^{10} 4s^2 4p^6 4d^{10} 5s$$

As physicists, we ask, "Why?" Why would silver decide that filling the 4d subshell is better than following that chart and keeping the 5s subshell filled? The short answer is that nature tends to try and get to a configuration with the lowest overall energy.

A General Rule of Nature
A system always seeks to reach a state of lower energy.

As an analogy, think about a ball in a bowl. The ball will want to hang out at the bottom of the bowl, which is the spot of lowest energy. Because there are so many electrons in silver and so many interactions, the electron configuration that fills the 4d shell has less energy than if the 5s shell was filled.

Quick quiz:
Considering the silver atom, how many electrons are in the 3d subshell? The answer is in the footnotes.[4]

---

[4] 10

Here is the electron configuration for the ground state of europium, which has 63 electrons:

$$1s^2 2s^2 2p^6 3s^2 3p^6 3d^{10} 4s^2 4p^6 4d^{10} 5s^2 5p^6 6s^2 4f^7$$

From this electronic configuration we can see which shells are filled. After the 6s subshell fills up, the 4f subshell starts to fill until we run out of electrons. There are 7 electrons in the unfilled 4f subshell. Combined with the Coulomb interactions, the end result is a pretty complicated energy level diagram, see Figs. 8.1 and 8.2. This is why I'm glad to be an experimentalist. I don't have to try to calculate these levels, just measure them ☺.

Summary
The dominant interactions in atoms that determine the energy level locations are the Coulomb interaction and the way electrons fill shells.

8.3   Term Symbols

Reminder Definition
- **Torque:** A measure of how effective a force is in causing an object to rotate. Applying torque changes the angular momentum of an object. There is energy associated with that torque.

We spent almost all of Chap. 7 studying angular momentum and emphasizing the orientation of the vectors. The theory of electricity and magnetism reveals an important fact: because electrons and the nucleus have charge, and the electrons have angular momentum, everything in the atom exerts a torque on everything else.

A Brief Aside
The Coulomb interaction is used to describe the interaction between charged particles, resulting in an electric field between the charges. This approach ignores the motion associated with their angular momentum. When we include angular momentum, it is more appropriate to use the electromagnetic interaction because moving charges create magnetic fields. Therefore, the interaction between moving charged particles involves both electric and magnetic forces, which is what the electromagnetic force describes.

Definitions
- **Electromagnetic force:** A fundamental force encompassing both electric and magnetic forces, including the Coulomb force and magnetic forces due to moving charges.
- **Electromagnetic interaction:** The interaction between charged particles due to both electric and magnetic forces, including interactions involving stationary charges (electrostatics) and moving charges.

The magnetic forces from the moving charges cause the internal torques. The magnitude of this internal torque depends on the orientation and size of the angular momentum vectors. In short, both the size and the orientation of the angular momentum vectors impact the energy of a state. That is so important, I'm going to give the sentence its own red box.

Important
Both the size and the orientation of the angular momentum vectors impact the energy of a state.

If we had a magic wand to change the orientation of an angular momentum vector, the energy of that state would change slightly. Now we can add a third property that affects where the energy levels end up: the Coulomb interaction, electron shells, and angular momentum.

The total orbital angular momentum of all the electrons, the total spin of all of the electrons, and total angular momentum of all of the electrons affects the energy of a state. Therefore, we need to include that information in the description of a state. By tradition, we combine these quantum numbers into a "term symbol", which looks like this:

$$^{2S+1}L_J \tag{8.1}$$

where S is the quantum number representing the total spin of all the electrons, L is the quantum number for the total orbital angular momentum of all the electrons, and J is the quantum number for the total angular momentum of all of the electrons. For example, suppose an energy level has a term symbol $^3D_2$. That tells us that the spin of the electrons has quantum number $S = 1$ ($2S + 1 = 3$), the orbital angular momentum of the electrons has quantum number $L = 2$ (represented by the letter D), and the total electronic angular momentum of the electrons has quantum number $J = 2$. We can use those numbers to calculate the magnitude of angular momentum.

Table 8.1 The first 6 states of atomic oxygen. The last column is the frequency of a laser needed if we were to try to excite the atom from the ground state to that state

Electron configuration	Term symbol	Energy (cm^{-1})	f (Hz)
$1s^2 2s^2 2p^4$	$^3P_2$	0	
$1s^2 2s^2 2p^4$	$^3P_1$	158.265	4.74×10^{12}
$1s^2 2s^2 2p^4$	$^3P_0$	226.977	6.80×10^{12}
$1s^2 2s^2 2p^4$	$^1D_2$	15,867.862	4.76×10^{14}
$1s^2 2s^2 2p^4$	$^1S_0$	33,792.583	1.01×10^{15}
$1s^2 2s^2 2p^3 3s$	$^5S_2$	73,768.200	2.21×10^{15}

Two Important Reminders

1. The term symbol is important not just because it tells us three quantum numbers for the electrons in an atom as a whole, but also because those three quantum numbers affect the energy of a state.
2. The observables represented by S, L, and J are all compatible with energy.

The shift of an energy level due to electrons having angular momentum is called fine structure splitting.[5] The nucleus can also have angular momentum (often called nuclear spin), which is the topic of Chap. 9. The shift in the energy of a state due to the nucleus having angular momentum is called hyperfine structure splitting.[6]

Definitions
- **Fine structure splitting:** The shift in the energy of a state due to the electrons having orbital angular momentum and spin.
- **Hyperfine structure splitting:** The shift in the energy of a state due to the nucleus having angular momentum.

Table 8.1 shows the six lowest energy states of atomic oxygen (8 electrons, 8 protons, and 8 neutrons). This particular type of oxygen is known as oxygen-16. Oxygen-16 has no nuclear angular momentum, so the nucleus does not play a role in the following discussion. Notice that the first five energy states all have the same electronic configuration but different term symbols. The next few paragraphs are all about adding lots and lots of angular momentum vectors together. We are going to start with the ground state and then work our way through the table.

---

[5] This is sometimes called fine splitting.

[6] This is sometimes called hyperfine splitting.

Reminder of Eq. 7.5

Suppose we have two quantum mechanical angular momenta vectors whose magnitudes are represented by quantum numbers L and S. Depending on the orientation of the two vectors, adding them together produces a new quantum mechanical angular momentum vector that can have different magnitudes, represented by the quantum numbers

$$J = (L + S), \ldots, |L - S|$$
in integer steps

For example, if $L = 3$ and $S = 1$, then J can be 4, 3, or 2. If $J = 4$, then there are 9 possible orientations of this new angular momentum vector represented by the quantum number $m_J = 4, 3, 2, 1, 0, -1, -2, -3,$ or -4. Similarly, if $J = 2$, then there are 5 possible orientations of this new angular momentum vector represented by the quantum number $m_J = 2, 1, 0, -1,$ or -2.

The ground state of oxygen has the designation $1s^2 2s^2 2p^4\ ^3P_2$. This designation tells us that both the 1s and 2s subshells are completely filled, and we can essentially ignore them. The 2p subshell is partially filled with 4 electrons (the 2p subshell can hold up to 6 electrons). From the electronic configuration, we know the orbital angular momentum of each electron. The 4 electrons in the 2p subshell each have $\ell = 1$, or each electron in this subshell has orbital angular momentum with size $\sqrt{\ell(\ell + 1)}\hbar = \sqrt{2}\hbar$.

Each of the electrons in the 2p subshell have the same magnitude of orbital angular momentum ($\ell = 1$), but have different orientations. If we were to add all four of those vectors together, we will get a new vector that represents the orbital angular momentum of all four electrons as a composite system. For the ground state, the orbital angular momentum of the composite system has a magnitude represented by quantum number $L = 1$ (P in the term symbol).

Those same four electrons also have individual spin vectors that add up to $S = 1$ ($2S + 1 = 3$). Again, that spin vector represents the total spin of all four electrons as a composite system. Since the orbital angular momentum vector for all four electrons has a size and orientation and the spin vector for all four electrons has a size and orientation, we can ask the question, "What are the possible magnitudes for the total electronic angular momentum for the composite system?" The answer is $J = 2, 1,$ or 0. The orientations that produce a vector whose magnitude is represented by $J = 2$ has the lowest energy, so that is the ground state.

Important Foreshadowing Statement
The total electronic angular momentum for the ground state ($J = 2$) has 5 possible orientations that, in the absence of nuclear spin, all have the same energy. In other words, there are 5 states that all have the same energy. When there are states that have the same energy, we call those states **degenerate**.

Now let's move on to the next energy level. This level has an energy that is $158.265\,\text{cm}^{-1}$ larger than the ground state. The only difference in the designation for the ground state and this state is that J in the term symbol ($J = 2$ for the ground state and $J = 1$ for this state). The orbital angular momentum for all four electrons and the spin for all four electrons still add together, but the result has a different total electronic angular momentum for the composite system. The orientation of all the electrons to produce this new composite vector has a different internal torque, so that orientation has a different energy than the $J = 2$ ground state.

The first 5 energy levels of atomic oxygen have the same electron configuration: $1s^2 2s^2 2p^4$. To determine how the angular momenta of the last 4 electrons combine, we look at the term symbols. As you can see from the table, there are multiple ways this can happen, each resulting in a different energy level. In a hypothetical world without the internal torque, all 5 of these levels would be degenerate. However, due to the internal interactions, these levels split into the 5 distinct levels observed experimentally. This phenomenon is known as fine structure splitting.[7]

While the nucleus of oxygen-16 has no angular momentum, a different nucleus might. That nuclear angular momentum, whose magnitude is represented by the quantum number I, exerts an additional internal torque that, once again, shifts the energy of the states. This additional internal torque "breaks" the degeneracy of the fine structure states. For example, oxygen-17 has nuclear spin $I = 5/2$ (6 possible orientations). The ground state has a total electronic angular momentum of $J = 2$ (5 possible orientations). Adding together vector J and vector I produces a new angular momentum vector whose magnitude is represented by the quantum number F with possible values that range from $J + I$ to $|J - I|$ in integer steps, or in this case $F = 9/2, 7/2, 5/2, 3/2$, and $1/2$. So the ground state of oxygen-17 further splits into six hyperfine levels, each with a different energy. From the "important foreshadowing statement," the orientation of a particular F vector will not change the energy. If $F = 3/2$, there are four orientations represented by $m_F = 3/2, 1/2, -1/2$ and -1. All four of these states are degenerate. We will explore hyperfine structure and how the nucleus affects our energy levels further in Chap. 9.

---

[7] For completeness, Einstein's theory of special relativity, which deals with the physics of fast-moving objects, also shifts the energy of a state and is included as part of fine structure splitting, but we will not explore that here. If this book has inspired you to pursue further studies in quantum mechanics ☺, you will learn about perturbation theory and the effects of electron velocity.

Common Question
If $J = 2$, there are 5 states with different orientations that have the same energy. Can we do anything to "split" those final orientations so they have different energies? Yes! The internal torque doesn't do it, but we can apply a torque from the outside to "break the degeneracy". Applying an external torque using an external magnetic field is called the Zeeman Effect, named after the Dutch physicist Pieter Zeeman. Applying an external torque by applying an electric field is called the Stark Effect, named after the German physicist Johannes Stark. Both of these effects "break" the degeneracy of those states.

Summary
The orientation and size of angular momentum impact the energy of a state. For an atom with no nuclear spin, a state can be described by the individual electron configurations, which tells us the angular momentum of each individual electron, and the term symbol, which tells us about the orientation of the angular momentum vectors of the electrons as a whole.

Super Short Summary
The orientation and size of the angular momentum vectors matter.

8.4   Connecting Angular Momentum to Orbitals

You may have learned about orbitals in high school chemistry. Orbitals are visual representations of different energy states for an electron in a hydrogen atom. Figures 8.6 and 8.7 show some orbital pictures for different energy states in hydrogen. These are analogous to the shaking energy modes we studied in Chap. 1 (Fig. 1.7) and Chap. 6 (Fig. 6.2). There are 3 numbers on each plot given in bra-ket notation. The first number is the principal quantum number (also called the shell number). The second is the orbital angular momentum quantum number, and the last is the projection of the orbital angular momentum quantum number along the z-axis (orientation). For example, $|4\ 3\ \text{-}1\rangle$ means $n = 4$, $\ell = 3$, and $m_\ell = -1$. You can think of all of these orbitals as different standing waves of the electron. The "loops" are a bit harder to see in 3-dimensional space, so in Fig. 8.6 we plot a few different orbitals in 3D and in Fig. 8.7 we plot a few cross sectional views. The reason they are so hard to visualize is that we only have 3 dimensions to view 4 dimensions of information.

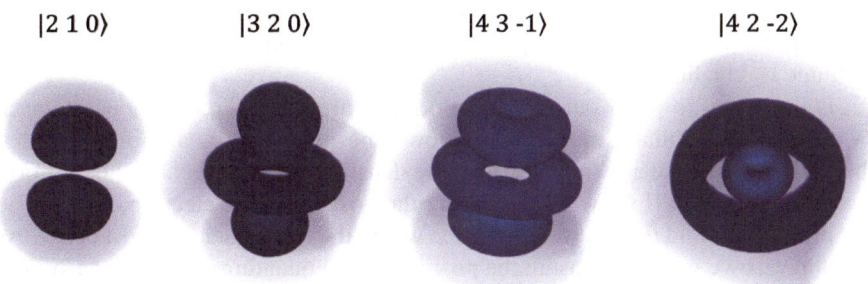

Fig. 8.6 3D illustrations of different electron orbitals in a hydrogen atom. The darker the shade the larger the amplitude of the standing wave

For example, if you look at Fig. 6.2 you'll see that we need two dimensions to see the energy state for 1 dimension of shaking energy. The vertical axis shows the amplitude of the energy state while the horizontal shows position in 1 dimension. An example of a 2 dimensional standing wave would be a drum head vibrating. To visualize a 2 dimensional standing wave, we need a 3 dimensional plot: 1 dimension for the amplitude and 2 for the position dimensions. We run into a problem for

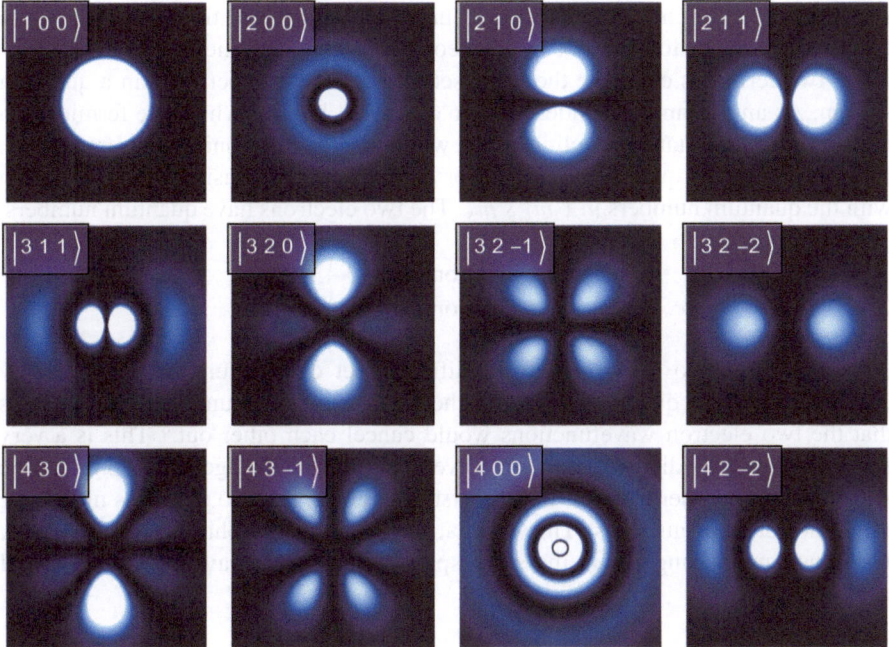

Fig. 8.7 A cross sectional view of different electron orbitals in the Hydrogen atom. The closer the color is to white, the larger the amplitude of the standing wave

3 dimensional energy states. We now need to plot 3 dimensions of position, but we also need to visualize the amplitude. Figures 8.6 and 8.7 attempt to show the amplitude by shading.

It is important to point out that these orbitals are not discrete regions of space (for example an orbital is not like a hollow ping pong ball or a uniformly filled sphere). It is more like a foam ball where the density of the foam ball is not uniform but varying in space. If we were to measure the position of the electron, it would be most probable to find the electron at a place where the energy state is most dense. As an analogy, if we were to measure the position of the quantum particle for a system in an energy state seen in Fig. 6.2, the most probable place to find the quantum particle would be where the amplitude is largest.

8.5 Fermions and Bosons

In atomic and nuclear physics, particles are categorized into two major groups: **fermions** and **bosons**. Fermions are particles with half-integer spin quantum numbers (e.g., 1/2, 3/2, etc.). Examples of fermions are electrons, protons, and neutrons, all of which have a spin quantum number of 1/2. Bosons are particles with integer spin quantum numbers (e.g., 0, 1, etc.). So far, we have encountered only one boson, the photon, which has a spin quantum number of 1.

This is important because two *identical* fermions cannot "hang out" in the same space while two identical bosons can. This is the essence of the **Pauli Exclusion Principle**: two fermions cannot simultaneously occupy the same quantum state; that is, no two fermions can have the same set of quantum numbers within a quantum system. As an example, consider the two electrons, both of which are fermions, in the atomic ground state of a helium atom, which has the electronic configuration $1s^2$. Let's write out the quantum numbers for the two electrons using bra-ket notation with the quantum numbers $|n \ \ell \ m_\ell \ s \ m_s$. The two electrons have quantum numbers:

$$\text{First Electron:} \quad |1 \ 0 \ 0 \ \tfrac{1}{2} \ \tfrac{1}{2}\rangle$$
$$\text{Second Electron:} \quad |1 \ 0 \ 0 \ \tfrac{1}{2} \ \text{-}\tfrac{1}{2}\rangle$$

Notice that the two electrons have a different set of quantum numbers. If they did have the same quantum numbers, the math from quantum mechanics shows that the two electron wavefunctions would cancel each other out.[8] This is a very bad thing. If the individual electron wavefunctions added together to produce no wavefunction, neither electron would exist. Therefore, the two electrons must have different quantum numbers. This is what is meant by the phrase, "two *identical* fermions cannot hang out in the same space". They must have a different set of quantum numbers.

---

[8] More specifically, the wavefunction for the two-electron system is "antisymmetric" and would be zero everywhere if the electrons had identical quantum numbers.

The Pauli Exclusion Principle is part of why electron shells and subshells exist. Every electron in the atom *must* have a different set of quantum numbers. Interestingly, bosons don't have this problem. Two bosons can have the same set of quantum numbers and not destructively interfere each other out of existence. So, two identical bosons can hang out in the same space. In fact, the two bosons can constructively interfere with each other.

There can also be composite fermions and bosons. For example, helium-4 is a system that acts as a composite boson. Helium-4 contains 2 protons, 2 neutrons, and 2 electrons. All of these particles are fermions, but they can pair up to behave like bosons. Other examples of composite bosons include Cooper pairs[9] (important for superconductors) and Bose–Einstein condensates[10]. Helium-3, which has 2 protons, 2 electrons, and 1 neutron, is a composite fermion.

Problems

8.1 As discussed in the chapter, hydrogen has 1 Coulomb interaction and helium has 3. Assuming the nucleus is one big particle with a positive charge, how many Coulomb interactions do lithium, beryllium, and boron have?

Challenge: What is the general formula to calculate the number of Coulomb interactions for an atom with N electrons, assuming the nucleus is one big particle with positive charge?

8.2 The electron configuration for hydrogen is 1s. There is 1 electron in the first s subshell. The electron configuration for helium is $1s^2$. That means there are 2 electrons in the first s subshell, which completely fills the shell. Lithium has 3 electrons, so the electron configuration is $1s^2 2s$. Notice since the first shell is filled, we begin to fill the second shell. Write out the electron configurations for

(a) beryllium (4 electrons)
(b) boron (5 electrons)
(c) carbon (6 electrons)
(d) nitrogen (7 electrons)
(e) oxygen (8 electrons)
(f) fluorine (9 electrons)
(g) neon (10 electrons)
(h) sodium (11 electrons).

---

[9] Also known as Bardeen–Cooper–Schrieffer pairs named after American physicists John Bardeen, Leon Cooper, and John Schrieffer.

[10] Named after Indian physicist Satyendra Nath Bose and German born physicist Albert Einstein.

Note: The electron configurations are something that you can easily find on the internet. You can also find the answers in Appendix B. Don't search for the answer before you try yourself first.

8.3 For each of the following term symbols, what is the magnitude of the total electronic spin, the total electronic orbital angular momentum, and the total electronic angular momentum?

(a) $^3P_2$
(b) $^3P_0$
(c) $^1S_0$
(d) $^1F_2$

8.4 Appendix B has a list of all of the elements with their ground state electronic configurations. Look through the list and find all the atoms that will have an energy level structure similar to hydrogen (i.e. all the shells are filled except for a single electron in the last s subshell). All of these elements will have energy level diagrams that look similar to hydrogen.

8.5 If we ionized (removed 1 electron) from beryllium, its electronic structure would look just like lithium, see Fig. 8.8. While the energy level spacings are proportionally similar, lithium is more compact. Why?

8.6 In Sect. 8.3 when exploring oxygen, we had the sentence, "This designation tells us that both the 1s and 2s subshells are completely filled, and we can essentially ignore them." Let's explore this sentence a bit more.

Fig. 8.8 Energy levels for a neutral lithium atom and a singly ionized beryllium atom

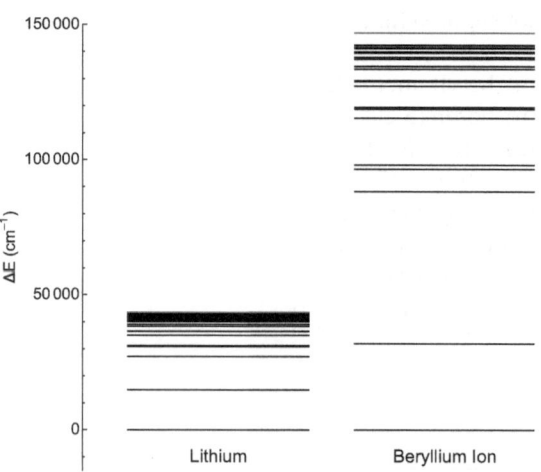

(a) Write the term symbol for the two electrons that fill the 1s subshell. Explain the reasoning you used to determine the term symbol. The answer (without explanation) is in the footnotes.[11]
(b) Write the term symbol for the two electrons that fill the 2s subshell. Explain your reasoning.
(c) Write the term symbol for the four electrons with the electronic configuration $1s^2 2s^2$. This is the same as the ground state of a beryllium atom, which has 4 electrons. Explain your reasoning.
(d) Write the term symbol for the ground state of neon. Neon has 10 electrons, so the 2p subshell is completely filled. Explain your reasoning.
(e) Write the term symbol for the ground state of lithium, which has 3 electrons. Explain your reasoning.
(f) Write the term symbol for the ground state of cesium, which has all subshells filled except for the last electron that is in the 6s subshell. Explain your reasoning.
(g) Boron has all subshells filled except for the last electron that is in the 2p subshell. For that last electron, there are two possible values for J. What are they?
(h) For part (g), the term symbol with the lower value of J corresponds to the lower energy and thus the ground state. Write the term symbol for the ground state of boron.

8.7

(a) Make a sketch of a saturated absorption spectroscopy spectrum where the lower state has $J = 0$ and the upper state has $J = 1$. The transition frequency is $f_r = 652,000,000$ MHz and the natural linewidth is 1 MHz.

Next, we put the atoms in an external magnetic field. The magnetic field shifts the energy levels according to the formula $\Delta E = \left(1\frac{\text{GHz}}{\text{T}}\right)h\, m_J\, B_{\text{ext}}$, where B_{ext} is the external magnetic field and the unit T stands for Tesla (the unit for magnetic field).

(b) We place the atoms in a uniform magnetic field with $B_{\text{ext}} = 0.1$ T. Make a sketch of a saturated absorption spectroscopy spectrum where the lower state has $J = 0$ and the upper state has $J = 1$. Don't worry about the relative amplitudes.
(c) Optional Challenge: Next, we place the atoms in a magnetic field described by the function $B_{\text{ext}} = \left(0.001\frac{\text{T}}{\text{cm}}\right)z$. Make a sketch of the three $J = 1$ states as a function of z. Use frequency units instead of Joules for the vertical axis (this is equivalent to using a laser to excite the atoms).

---

[11] $^1S_0$.

Hyperfine Structure

9

Abstract

In this chapter, we explore hyperfine structure, which occurs in atoms with a nucleus that has angular momentum. We will discuss the theoretical framework, including key equations for calculating hyperfine energy shifts, and apply this knowledge through practical examples involving cesium, europium, and oxygen. This chapter aims to provide a comprehensive understanding of hyperfine structure and its significance in atomic physics.

Learning Goals

By the end of this chapter, you should be able to understand:

- why hyperfine structure exists and identify the conditions under which hyperfine splitting occurs.
- the difference between fine structure and hyperfine structure.
- how to interpret spectroscopic data to extract hyperfine constants.

9.1   Hyperfine Structure

The nucleus contains protons and neutrons, each of which has intrinsic angular momentum. In addition to their intrinsic spins, nucleons (protons and neutrons) can also have orbital angular momentum due to their motion within the nucleus. When discussing the angular momentum of the nucleus as a whole, we primarily refer to the combination of the intrinsic spins of the nucleons and their orbital angular momentum. This combined angular momentum should be called the "total nuclear angular momentum," but it is commonly referred to as "**nuclear spin**." Despite the

© The Author(s) 2025
W. Raven, *Atomic Physics for Everyone*,
https://doi.org/10.1007/978-3-031-69507-0_9

terminology, neither the nucleus nor the protons and neutrons in the nucleus are literally spinning or orbiting in the classical sense. Nuclei with both an even number of protons and an even number of neutrons tend to have zero nuclear spin because the individual spins pair up (one spin-up and one spin-down) and cancel each other out. Nuclei with an odd number of protons and/or neutrons generally have a non-zero nuclear spin.

When an atom has a non-zero nuclear spin, the nucleus interacts with the magnetic fields produced by the electrons. More specifically, because the nucleus has both charge and angular momentum (I), and the electrons have both charge and angular momentum (J), there is an additional internal torque between the nucleus and the electrons. This interaction leads to the splitting of atomic energy levels into closely spaced sub-levels known as hyperfine levels, analogous to the interaction between an electron's spin and its orbital angular momentum, which results in fine structure splitting. I want to point out that atomic physicists use **hyperfine levels** and hyperfine states interchangeably. We tend to use hyperfine levels when thinking about energy splittings and hyperfine states when using quantum numbers, but the two terms mean the same thing.

Table 9.1 is a copy of a table from Chap. 7 that summarizes all of the angular momentum vectors and angular momentum quantum numbers for the system of electrons and the atom as a whole. Let's look at some examples. Oxygen has three stable isotopes: oxygen-16 (99.76% of all oxygen on earth is oxygen-16), oxygen-17 ($\sim$0.04%), and oxygen-18 ($\sim$0.20%). Isotopes are elements with the same number of protons but different numbers of neutrons. Oxygen-16 has 8 electrons, 8 protons, and 8 neutrons. Oxygen-17 has 8 electrons, 8 protons, and 9 neutrons. Oxygen-18 has 8 electrons, 8 protons, and 10 neutrons. Since each isotope has a different number of neutrons, the transition frequencies are slightly different. This small shift, called an isotope shift, will be discussed in Chap. 10.

Table 9.1 This table summaries all of the angular momentum quantum numbers

Type	QN	Rule	Formula
Orbital	L	Zero or positive integer	Magnitude: $\sqrt{L(L+1)}\hbar$
	m_L	$-L$ to $+L$ in integer steps	Cone height: $m_L\hbar$
Spin	S	Zero, positive integer, or half-integer	Magnitude: $\sqrt{S(S+1)}\hbar$
	m_S	$-S$ to $+S$ in integer steps	Cone height: $m_S\hbar$
Total electronic	J	Zero, positive integer, or half-integer	Magnitude: $\sqrt{J(J+1)}\hbar$
	m_J	$-J$ to $+J$ in integer steps	Cone height: $m_J\hbar$
Nuclear spin	I	Zero, positive integer, or half-integer	Magnitude: $\sqrt{I(I+1)}\hbar$
	m_I	$-I$ to $+I$ in integer steps	Cone height: $m_I\hbar$
Total atomic	F	Zero, positive integer, or half-integer	Magnitude: $\sqrt{F(F+1)}\hbar$
	m_F	$-F$ to $+F$ in integer steps	Cone height: $m_F\hbar$

Fig. 9.1 The hyperfine
structure of the ground state
of oxygen-17. The hyperfine
levels are shown with respect
to the center of gravity of the
ground state

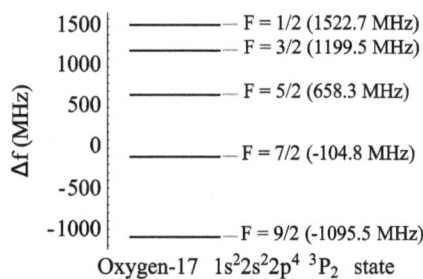

Oxygen-17 $1s^2 2s^2 2p^4\ {}^3P_2$ state

Important Reminders

- We can represent the energy difference between two states using energy
  units, wavelength units, or frequency units. For hyperfine structure, frequency units are the most convenient unit because the energy spacing
  between hyperfine levels is small.
- When adding two angular momentum vectors together, both the magnitude
  and orientation of the individual angular momentum vectors will affect the
  magnitude and orientation of the resulting vector. Each possible magnitude
  of the composite angular momentum vector will have an effect on the
  energy of a state.
- **Center of gravity:** The energy of a state if there was no nuclear spin.

The nuclear spin quantum number for oxygen-17 is $I = 5/2$. The nuclear spin
quantum number for both oxygen-16 and oxygen-18 is $I = 0$. Therefore oxygen-17
will have hyperfine structure while the other two isotopes do not. All isotopes of
oxygen have a ground state term symbol of ${}^3P_2$, or $S = 1$ ($2S + 1 = 3$), $L = 1$,
and $J = 2$. However, the ground state of oxygen-17 looks different compared to
the other two isotopes, see Fig. 9.1. If oxygen-17 had no nuclear spin, it would have
a single level precisely at 0. Nuclear spin "splits" this single level into 5 hyperfine
levels. For example, the level labeled $F = 7/2$ has a slightly smaller energy than
the center of gravity while the $F = 5/2$ state has a higher energy. Both oxygen-16
and oxygen-18 have no nuclear spin, so they have a single ground state that would
be labeled 0 energy and have no F quantum number designation.

Adding together nuclear spin, represented by the quantum number I, and the total
electronic angular momentum, represented by the quantum number J, results in a
new angular momentum vector that we call the total angular momentum of the atom,
represented by the quantum number F. Hyperfine structure generally has a much
smaller energy splitting compared to fine structure. Like every angular momentum
vector we have encountered, the F vector can also point in different orientations. For
example, an $F = 3/2$ state has four possible orientations described by the quantum

numbers $m_F = 3/2$, $1/2$, $-1/2$, and $-3/2$. All four of those orientations have the same energy.

Compatibility with Energy We often care about what observables are compatible with energy. At the end of Chap. 7, we used the ket $|n \ \ell \ s \ j \ m_j\rangle$ to represent the states of the hydrogen atom (one electron). All the different quantum numbers in this ket represent observables that are compatible with energy.

> **Important Reminder**
> The observables represented by m_ℓ and m_s are not compatible energy.

Let's update our ket with information we have learned from this chapter. When we add together nuclear spin and total electronic angular momentum, the cones representing those angular momenta are tilted with respect to the total atomic angular momentum (F). Just like when we added orbital and electron spin, the cone heights of I and J, which are represented by the quantum numbers m_I and m_J, are no longer compatible with energy. Therefore, our new ket is $|n \ \ell \ s \ j \ I \ F \ m_F\rangle$.

$$\begin{aligned} &|n \ \ell \ s \ j \ m_j\rangle && \text{Energy state with no nuclear spin} \\ &|n \ \ell \ s \ j \ I \ F \ m_F\rangle && \text{Energy state with nuclear spin} \end{aligned} \tag{9.1}$$

The nucleus of a hydrogen atom has a nuclear spin represented by $I = 1/2$. The atomic ground state of hydrogen has the electronic configuration 1s, or $s = 1/2$, $\ell = 0$, and $j = 1/2$. Because the nucleus has spin, there will be two hyperfine levels represented by $F = 0$ ($m_F = 0$) and $F = 1$ ($m_F = 1, 0$, and -1). Therefore, the four hyperfine levels for the ground state have kets:

$$\begin{aligned} &|1 \ 0 \ \tfrac{1}{2} \ \tfrac{1}{2} \ \tfrac{1}{2} \ 1 \ 1\rangle \\ &|1 \ 0 \ \tfrac{1}{2} \ \tfrac{1}{2} \ \tfrac{1}{2} \ 1 \ 0\rangle \\ &|1 \ 0 \ \tfrac{1}{2} \ \tfrac{1}{2} \ \tfrac{1}{2} \ 1 \ \text{-}1\rangle \\ &|1 \ 0 \ \tfrac{1}{2} \ \tfrac{1}{2} \ \tfrac{1}{2} \ 0 \ 0\rangle \end{aligned} \tag{9.2}$$

The three hyperfine levels with $F = 1$ are degenerate and have the same energy. The hyperfine level with $F = 0$ has a different energy.

So, the ground state of hydrogen has two hyperfine levels represented by the quantum numbers $F = 1$ and $F = 0$. For something a bit more complicated, let's look at the ground state of europium. Europium has 7 electrons in its last, unfilled subshell. All isotopes of europium have a ground state term symbol $^8S_{7/2}$, or $S = 7/2$ ($2S + 1 = 8$), $L = 0$, and $J = 7/2$. If europium had no nuclear spin, that would be the end of the story. We would define the ground state:

$$1s^2 2s^2 2p^6 3s^2 3p^6 3d^{10} 4s^2 4p^6 4d^{10} 5s^2 5p^6 6s^2 4f^7 \ ^8S_{7/2} \tag{9.3}$$

to have 0 energy. However, both stable isotopes of europium have nuclear spin: europium-151 (63 electrons, 63 protons, 88 neutrons, and $I = 5/2$) and europium-153 (63 electrons, 63 protons, 90 neutrons, and $I = 5/2$). Other isotopes of europium will have different nuclear spin. For example, europium-152, which is radioactive with a half-life of 13.5 years, has a nuclear spin quantum number of $I = 3$. Each of these europium isotopes have hyperfine levels.

Summary
If an isotope has no nuclear spin, there would be 1 ground state. If it does have nuclear spin, there are multiple ground states. This is because the angular momentum from the nucleus exerts a torque on the electrons, which results in a small splitting and shift of the state's energy.

There is a single exception to the above summary that we will discuss more in Sect. 9.2. If the state has no total electronic angular momentum ($J = 0$), there is still only a single state; the single state does not split into multiple hyperfine levels. However, that single state will still have an F quantum number as a label.

9.2   Math

We can find the possible values of F by using the largest to smallest in integer steps rule. The rule is:

$$F = (I + J), \ldots, |I - J| \quad \text{in integer steps.} \tag{9.4}$$

Let's do a few examples.

Example 1 Hydrogen-1 has a nuclear spin quantum number of $I = 1/2$ and the ground state has a total electronic angular momentum quantum number of $J = 1/2$ (also $j = 1/2$ since hydrogen has a single electron). Therefore, the possible F values for the ground state range from $1/2 + 1/2 = 1$ to $|1/2 - 1/2| = 0$ in integer steps. Thus, there are 2 hyperfine levels for the ground state of hydrogen-1.

Example 2 Oxygen-17 has a nuclear spin quantum number of $I = 5/2$ and the ground state has a total electronic angular momentum quantum number of $J = 2$. Therefore, the possible F values for the ground state range from $5/2 + 2 = 9/2$ to $|5/2 - 2| = 1/2$ in integer steps. Thus, there are 5 hyperfine levels for the ground state of oxygen-17. Those values, which are shown in Fig. 9.1, are $F = 9/2, 7/2,$ 5/2, 3/2, and 1/2.

Example 3 Let's consider the two stable isotopes of europium: europium-151 and europium-153. Both stable isotopes of europium have the same nuclear spin quantum number of $I = 5/2$. The ground state has a total electronic angular momentum quantum number of $J = 7/2$. Therefore, the possible values for F range from $5/2 + 7/2 = 6$ to $|5/2 - 7/2| = 1$ in integer steps, resulting in $F = 6, 5,$ 4, 3, 2, or 1.

Quick Quiz

1. The radioactive isotope europium-152 has a nuclear spin quantum number of $I = 3$. How many hyperfine levels will the ground state have, and what are their F quantum numbers?
2. Beryllium-9 has a nuclear spin quantum number of $I = 3/2$ with a ground state total electronic angular momentum quantum number of $J = 0$. How many hyperfine levels will the ground state have, and what are their F quantum numbers? The answers are below the maze shown in Fig. 9.2.

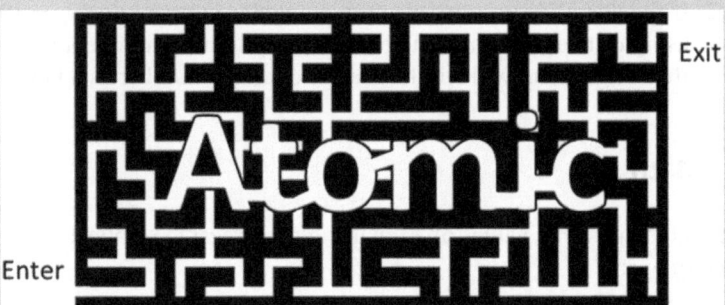

Fig. 9.2 A fun maze to separate the quiz from the answers

The answers are:

1. 7 levels: $F = 13/2, 11/2, 9/2, 7/2, 5/2, 3/2,$ or $1/2$
2. 1 level: $F = 3/2.$

Notice that beryllium-9 still has a single ground state even though the nucleus has spin. This is because $(I + J) = |I - J|$, so the range of possible F values is $3/2$ to $3/2$. This always happens when $J = 0$. However, the nucleus still has angular momentum, so we include $F = 3/2$ in the description of the state.

Using quantum mechanics, we can derive the energy splitting of a hyperfine level with respect to the center of gravity:

$$\Delta E = \tfrac{1}{2}KA + \frac{\tfrac{3}{2}K(K+1) - 2I(I+1)J(J+1)}{2I(2I-1)2J(2J-1)}B$$
$$K = F(F+1) - I(I+1) - J(J+1)$$
$$A = 0 \text{ unless both } I > 0 \text{ and } J > 0$$
$$B = 0 \text{ unless both } I > 1/2 \text{ and } J > 1/2$$

(9.5)

where A is called the magnetic dipole hyperfine constant and B is called the electric quadrupole hyperfine constant. Notice that everything else in the above equation apart from A and B is a quantum number. Theorists can calculate the hyperfine constants A and B while experimentalists measure them. When we perform spectroscopy on an atom with hyperfine structure, we can measure the energy spacing between all of the hyperfine levels and use the above equation to back out experimental values of A and B. Some hyperfine constants were measured many years ago while others have yet to be measured. For example, the ground state hyperfine constants for europium-151 and europium-153 were measured for the first-time way back in 1960 by P.G.H. Sandars and G.K. Woodgate and published in the journal *Proceedings of the Royal Society A*.[1] Sandars and Woodgate found that the magnetic dipole hyperfine constant and the electric quadrupole hyperfine constant for the ground state of europium-151 is $A = -20.0523 \pm 0.0002\,\text{MHz}$ and $B = -0.7012 \pm 0.0035\,\text{MHz}$. To convert those numbers into energy, just plug the hyperfine constants into the above equation and multiply the result by Planck's constant, h. If someone already measured those numbers, we can use those as a starting point for our fitting algorithms. If not, we have to determine them ourselves.

For every state in an atom, the electrons have different quantum numbers. States with higher n tend to be farther from the nucleus while the angular momentum quantum numbers represent different orbitals. Therefore, for an atom with nuclear spin, every state in that atom will have different hyperfine constants resulting in a different hyperfine splitting. Even the same state in two different isotopes that happen to have the same nuclear spin will have different hyperfine constants because the nuclei of the two isotopes are slightly different.

Summary
The "splitting" of the center of gravity into hyperfine levels is described by Eq. 9.5. The magnetic dipole hyperfine constant A is zero unless both $I > 0$ and $J > 0$. The magnetic quadrupole hyperfine constant B is zero unless both $I > 1/2$ and $J > 1/2$.

---

[1] How cool of a journal name is that?! See reference [2] for the full citation.

One Final Thing The magnetic dipole term (the term with A) in Eq. 9.5 tends to be larger than the electric quadrupole term (the term with B). For example, the $F = 6$ ground hyperfine state of europium-151 has a splitting $\Delta E = (-175.458\,\text{MHz}) + (-0.175\,\text{MHz}) = -175.633\,\text{MHz}$. The first term in parentheses is from the magnetic dipole term and the second is from electric quadrupole term. There are additional terms to Eq. 9.5, but they are very small compared to the electric quadrupole term. The next term in the formula is the magnetic octupole term, which contains quantum numbers and the magnetic octupole hyperfine constant C. This term is generally unnecessary unless you have exceptionally good data. The magnetic octupole constant is zero unless both $I > 1$ and $J > 1$.

9.3 Transition Frequencies

Suppose we have an atom with a nuclear spin quantum number of $I = 3/2$. To help distinguish between the lower and upper states, we will use primes on the quantum numbers for the excited states. In this example, the lower state has a total angular momentum quantum number of $J = 1/2$ and the upper state has $J' = 1/2$. Our goal is to write an equation for the transition frequency between two hyperfine levels. We first need to find the possible values for F, which can range from $3/2 + 1/2 = 2$ to $|3/2 - 1/2| = 1$ in integer steps, giving $F = 1$ and $F = 2$. Since $J' = 1/2$ as well, the possible values for F' are $F' = 1$ and $F' = 2$, see Fig. 9.3. In this hypothetical example, the $F = 1$ hyperfine level has a smaller energy than the $F = 2$ hyperfine level while the order is reversed in the excited state; the ordering of the quantum number all depends upon the interaction with the nucleus.

Next, we want to find the transition frequency from the $F = 2$ hyperfine level to the $F' = 1$ hyperfine level. Using Eq. 9.5, we can calculate the hyperfine energy splitting. We will use "LS" for lower state and "US" for upper state. Note that since $J = 1/2$ and $J' = 1/2$, both $B_{\text{LS}} = 0$ and $B_{\text{US}} = 0$. Evaluating Eq. 9.5 with the given quantum numbers, we find $\Delta E_{\text{LS},F=2} = \frac{3}{4}A_{\text{LS}}$ and $\Delta E_{\text{US},F'=1} = -\frac{5}{4}A_{\text{US}}$. According to Fig. 9.3, the $F = 2$ state has more energy than the center of gravity for the lower level, meaning $\Delta E_{\text{LS},F=2} > 0$. Since $\Delta E_{\text{LS},F=2} = \frac{3}{4}A_{\text{LS}}$, we also learn that $A_{\text{LS}} > 0$. For the upper state, the hyperfine energy splitting is also positive.

Fig. 9.3 A simple Grotrian diagram for a made-up atom

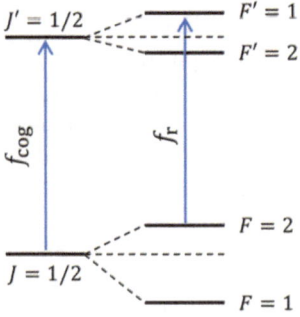

For this state, the formula is $\Delta E_{\text{US},F'=1} = -\frac{5}{4}A_{\text{US}}$, which implies that $A_{\text{US}} < 0$ in order to make $\Delta E_{\text{US},F'=1} > 0$.

Quick Quiz

Suppose an electron in our made up atom is in the $F = 2$ lower hyperfine level. What is the formula to calculate the transition frequency from the $F = 2$ lower hyperfine level to the $F' = 1$ upper hyperfine level? Your answer should contain the center of gravity frequency f_{cog} and the two magnetic dipole hyperfine constants A_{LS} and A_{US}. The answer is in the footnotes.[2]

The general formula to calculate the transition frequency between two hyperfine levels is

$$f_r = f_{\text{cog}} - \Delta E_{\text{LS}}(A_{\text{LS}}, B_{\text{LS}}) + \Delta E_{\text{US}}(A_{\text{US}}, B_{\text{US}}). \tag{9.6}$$

For many learners, the signs of the shifts can be confusing, so let's explore the signs with our toy example. Let's start with the minus sign in front of $\Delta E_{\text{LS}}(A_{\text{LS}}, B_{\text{LS}})$. According to Fig. 9.3, the $F = 2$ hyperfine state reduces the transition frequency compared to f_{cog}. Since $\Delta E_{\text{LS}}(A_{\text{LS}}, B_{\text{LS}}) > 0$ for that state, the minus sign makes sense. What about if we wanted the transition frequency from the $F = 1$ lower hyperfine level? For that state, $\Delta E_{\text{LS}}(A_{\text{LS}}, B_{\text{LS}}) < 0$, so that minus sign turns into a plus sign, increasing the frequency needed compared to f_{cog}. This is a subtle but important point, so take some time to convince yourself the transition frequency from any lower hyperfine level to any upper hyperfine level is given by Eq. 9.6. Using this same logic, convince yourself that the plus sign in front of $\Delta E_{\text{US}}(A_{\text{US}}, B_{\text{US}})$ makes sense.

9.4   Example with Cesium-133

Figure 9.4 shows a spectrum from a paper we published in 2018,[3] in which we performed saturated absorption spectroscopy on cesium-133. Cesium-133 has a nuclear spin quantum number of $I = 7/2$. The first five shells of cesium are completely filled, leaving a single electron in the 6s subshell. The ground state has the term symbol $^2S_{1/2}$ while the excited state has a term symbol $^2P_{3/2}$. So, the ground state has two hyperfine levels, $F = 3$ and $F = 4$, while the excited state has four hyperfine levels, $F' = 2, 3, 4,$ and 5. The $F = 3$ and $F = 4$ ground hyperfine levels are well separated from each other ($\sim 9.192\,\text{GHz}$), which is much larger than

---

[2] $f_{F=2 \to F'=1} = f_{\text{cog}} - \frac{3}{4}A_{\text{LS}} - \frac{5}{4}A_{\text{US}}$.

[3] Reference [1].

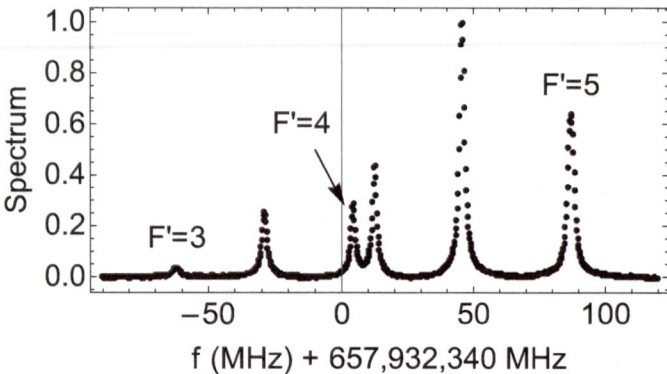

Fig. 9.4 Experimental spectroscopic results of the 6s $^2S_{1/2}$ $F = 4 \rightarrow$ 7p $^2P_{3/2}$ transition in neutral cesium-133. This is an experimental result from my research group, see reference [1]. A simplified Grotrian diagram for the transition can be found in Fig. 5.12

the Doppler width for transitions from either state. Therefore, we will not have any Λ crossovers despite having two lower states.

We performed spectroscopy from the $F = 4$ ground hyperfine level, which has an energy approximately 4.021 GHz above the center of gravity for the ground state. Given The Rule (Eq. 3.15) $\Delta F = -1, 0,$ or 1 with the exception $F = 0 \not\rightarrow F = 0$, we can excite an atom from the ground state hyperfine level with $F = 4$ to the excited state hyperfine levels with $F' =$3, 4, or 5.[4] Each of these three real transitions has a Lorentzian lineshape. We also have three V crossovers. For a review of crossovers, see Chap. 5, Sect. 5.2. The first crossover comes from the two real transitions $F = 4 \rightarrow F' = 3$ and $F = 4 \rightarrow F' = 4$. We often write this crossover in shorthand form: $F = 4 \rightarrow F' = 3/4$. The other two crossovers are the $F = 4 \rightarrow F' = 4/5$ and $F = 4 \rightarrow F' = 3/5$ crossovers. Despite having 6 spectroscopic features, the frequencies of all 6 features depend on only five parameters: the two hyperfine constants for the ground state, the two hyperfine constants for the excited state, and the transition frequency between the center of gravity for the ground state and the center of gravity for the excited state, see Eq. 9.6. I should point out that the width and amplitude of each peak are also free parameters, but we didn't really care about those. Our main goal was to measure the center of gravity frequency and the hyperfine constants. Having 5 parameters to determine the position of all spectroscopic features is typical. The experimental data from our group on a transition in europium-151 had up to 77 spectral features in that spectrum![5] Even still, the center of each feature is determined by only 5 free parameters.

---

[4] A full list of the rules that need to be satisfied for an electron to transition between two atomic states is given in Appendix C.

[5] See Reference [4] for the full citation.

Let's apply Eq. 9.6 using numbers from cesium-133. The spectrum in Fig. 9.4 is from the $J = 1/2$, $F = 4$ ground state. The excited state has $J' = 3/2$ and $F' = 3$, 4, and 5.[6] Let's find ΔE for all of the features.

9.4.1 Real Transitions

There are three real transitions in the spectrum. They are

1. $F = 4 \rightarrow F' = 3$
2. $F = 4 \rightarrow F' = 4$
3. $F = 4 \rightarrow F' = 5$

Let's calculate the shift from the center of gravity for the 7p $^2P_{3/2}$ state for each real transition. Since these are shifts for the excited state, we will use the primed values in Eq. 9.5. We will also need Eq. 9.6 to find the transition frequency between two hyperfine levels. Since the hyperfine splitting for the ground state is so large, we will perform our spectroscopy with respect to the $F = 4$ ground hyperfine level. For ease of reference, here are those equations:

$$\Delta E = \tfrac{1}{2}K A_{\text{US}} + \frac{\tfrac{3}{2}K(K+1)-2I(I+1)J'(J'+1)}{2I(2I-1)2J'(2J'-1)} B_{\text{US}}$$
$$K = F'(F'+1) - I(I+1) - J'(J'+1) \tag{9.7}$$

$$f_r = f_{F=4\rightarrow\text{cog}} + \Delta E_{\text{US}}(A_{\text{US}}, B_{\text{US}})$$

where $f_{F=4\rightarrow\text{cog}} = f_{\text{cog}} - \Delta E_{\text{LS,F=4}}$ is the frequency of light needed to go from the $F = 4$ ground hyperfine level to the center of gravity of the 7p $^2P_{3/2}$ state. I also replaced the quantum numbers with primed values and the hyperfine constants with "US" subscripts.

Let's do the math for the $F = 4 \rightarrow F' = 3$ real transition. First, let's find K:

$$\begin{aligned} K &= F'(F'+1) - I(I+1) - J'(J'+1) \\ &= 3(3+1) - \tfrac{7}{2}(\tfrac{7}{2}+1) - \tfrac{3}{2}(\tfrac{3}{2}+1) \\ &= -\tfrac{15}{2} \end{aligned}$$

Now we can find the hyperfine splitting for the $F' = 3$ state from the center of gravity for the 7p $^2P_{3/2}$ state.

---

[6] There is also an $F' = 2$ excited hyperfine level, but we cannot excite an atom from the $F = 4$ ground hyperfine level to this state.

Table 9.2 The energy shifts for the hyperfine levels with respect to the center of gravity for the $7p\,^2P_{3/2}$ state in cesium-133. All of the real transitions in this example are from the $6s\,^2S_{1/2}$ $F = 4$ ground hyperfine level

Transitions	$K = F'(F'+1) - I(I+1) - J'(J'+1)$	ΔE
$F = 4 \to F' = 3$	$3(3+1) - \frac{7}{2}(\frac{7}{2}+1) - \frac{3}{2}(\frac{3}{2}+1) = -\frac{15}{2}$	$-\frac{15}{4}A_{US} - \frac{5}{28}B_{US}$
$F = 4 \to F' = 4$	$4(4+1) - \frac{7}{2}(\frac{7}{2}+1) - \frac{3}{2}(\frac{3}{2}+1) = +\frac{1}{2}$	$+\frac{1}{4}A_{US} - \frac{13}{28}B_{US}$
$F = 4 \to F' = 5$	$5(5+1) - \frac{7}{2}(\frac{7}{2}+1) - \frac{3}{2}(\frac{3}{2}+1) = +\frac{21}{2}$	$+\frac{21}{4}A_{US} + \frac{1}{4}B_{US}$

$$
\begin{aligned}
\Delta E &= \tfrac{1}{2}K A_{US} + \frac{\tfrac{3}{2}K(K+1) - 2I(I+1)J'(J'+1)}{2I(2I-1)2J'(2J'-1)}B_{US} \\
&= \tfrac{1}{2}\left(-\tfrac{15}{2}\right)A_{US} + \frac{\tfrac{3}{2}\left(-\tfrac{15}{2}\right)\left(\left(-\tfrac{15}{2}\right)+1\right) - 2\left(\tfrac{7}{2}\right)\left(\tfrac{7}{2}+1\right)\left(\tfrac{3}{2}\right)\left(\tfrac{3}{2}+1\right)}{2\left(\tfrac{7}{2}\right)\left(2\left(\tfrac{7}{2}\right)-1\right)2\left(\tfrac{3}{2}\right)\left(2\left(\tfrac{3}{2}\right)-1\right)}B_{US} \\
&= -\tfrac{15}{4}A_{US} + \frac{\left(-\tfrac{45}{4}\right)\left(-\tfrac{13}{2}\right) - 7\left(\tfrac{9}{2}\right)\left(\tfrac{3}{2}\right)\left(\tfrac{5}{2}\right)}{7(6)(3)(2)}B_{US} \\
&= -\tfrac{15}{4}A_{US} + \frac{\left(\tfrac{585}{8}\right) - \left(\tfrac{945}{8}\right)}{252}B_{US} \\
&= -\tfrac{15}{4}A_{US} + \frac{\left(-\tfrac{360}{8}\right)}{252}B_{US} \\
&= -\tfrac{15}{4}A_{US} + \left(\tfrac{-45}{252}\right)B_{US} \\
&= -\tfrac{15}{4}A_{US} - \tfrac{5}{28}B_{US}
\end{aligned}
$$

Table 9.2 shows the results of the math. The hyperfine constants for the $7p\,^2P_{3/2}$ state are kept as unknowns that we find while fitting the data. If there weren't any crossovers, we would be done. Our fitting function would be the sum of three Lorentzian functions[7] whose centers are taken from the above table:

$$
g(f) = \frac{C_1}{1 + \frac{\left(f - (f_{F=4 \to cog} - \frac{15}{4}A_{US} - \frac{5}{28}B_{US})\right)^2}{\gamma_1^2}} + \frac{C_2}{1 + \frac{\left(f - (f_{F=4 \to cog} + \frac{1}{4}A_{US} - \frac{13}{28}B_{US})\right)^2}{\gamma_2^2}}
$$
$$
+ \frac{C_3}{1 + \frac{\left(f - (f_{F=4 \to cog} + \frac{21}{4}A_{US} + \frac{1}{4}B_{US})\right)^2}{\gamma_3^2}}. \tag{9.8}
$$

where $f_{F=4 \to cog}$ is the frequency of light needed to go from the $F = 4$ ground hyperfine level to the center of gravity of the $7p\,^2P_{3/2}$ state. To avoid accidentally mixing up the magnetic dipole hyperfine constant of the $7p\,^2P_{3/2}$ state and the amplitude of the Lorentzian functions, I changed the variable for amplitude to C.

---

[7] There are many subtleties that we are glossing over here. Spectra sometimes have an offset, a sloped offset, or a Gaussian pedestal for various reasons. When you fit data, sometimes you have to add something to your fit function. However, you should always have a reason for why you are adding something to your fit function.

If we wanted to write the transition frequencies in terms of the actual center of gravity frequency, we would replace $f_{F=4\to\text{cog}}$ with $f_{\text{cog}} - \Delta E_{LS,F=4}$.

9.4.2 Crossover Transitions

For this spectrum, there are three V crossovers: $F = 4 \to F' = 3/4$, $F = 4 \to F' = 4/5$, and $F = 4 \to F' = 3/5$. The center of the spectral feature due to the $F = 4 \to F' = 3/4$ crossover is found using the same procedures we learned about in Chap. 5, which is adding the two real transitions and dividing by 2:

$$\Delta E_{F'=3/4} = \frac{\Delta E_{F'=3} + \Delta E_{F'=4}}{2} = \frac{\left(-\frac{15}{4}A_{\text{US}} - \frac{5}{28}B_{\text{US}}\right) + \left(\frac{1}{4}A_{\text{US}} - \frac{13}{28}B_{\text{US}}\right)}{2}$$
$$= -\frac{7}{4}A_{\text{US}} - \frac{9}{28}B_{\text{US}} \qquad (9.9)$$

The same procedure can be done for the other two crossovers. In the end, the fit function is a sum of 6 Lorentzian functions. From the fit, we find A_{US}, B_{US}, and $f_{F=4\to\text{cog}}$, all with uncertainty. You will get to practice this in a homework problem.

9.5 Optional: Amplitudes

A common question I am asked is if there is a way to calculate the amplitudes of the Lorentzian functions. Finding the absolute amplitude is possible, but quite difficult. However, if the spectral features aren't overlapping (we often use the phrase "well-separated features"), we can find the relative amplitudes for the real transitions. It uses a mathematical symbol you may not have seen before, but you should think of it as a placeholder for a lot of hidden algebra. Not many people do the hidden algebra themselves—that's what Mathematica is for—but if you search the internet for "Wigner 6-j," you can find all the math behind the symbol. The scaled amplitudes of the real transitions are given by the following formula:

$$I_r = (2F + 1)(2F' + 1)\begin{Bmatrix} J & I & F \\ F' & 1 & J' \end{Bmatrix}^2, \qquad (9.10)$$

where the primed quantum numbers are for the excited state. That last symbol is called the Wigner 6-j symbol (it is squared in the above equation). Also, the second element of the second row is the number 1 (lots of folks accidentally read that as the nuclear spin quantum number I). In Mathematica, you would type SixJSymbol{J,I,F},{Fp,1,Jp}^{2} (Mathematica doesn't allow J' or F' as variable names, so I replaced them with Jp and Fp). There are also online calculators you can find by searching the internet for "Wigner 6-j symbol calculator". Notice this formula has no units. We use this formula to take ratios to find relative amplitudes.

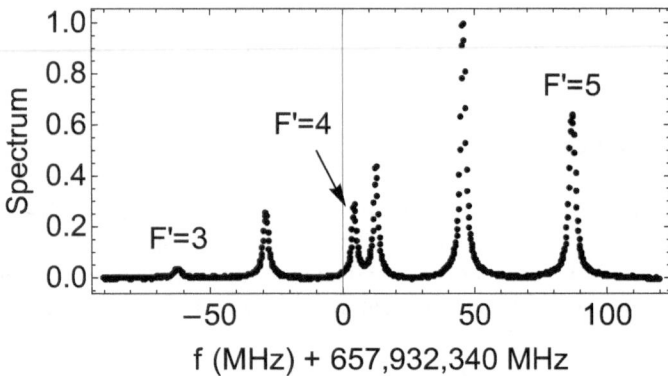

Fig. 9.5 Experimental spectroscopic results of the 6s $^2S_{1/2}$ $F = 4 \rightarrow$ 7p $^2P_{3/2}$ transition in neutral cesium-133. This is an experimental result from my research group. A simplified Grotrian diagram for the transition can be found in Fig. 5.12

Let's find I_r for all three real transitions in the cesium-133 example we have been studying. To make things a bit easier, I included a copy of Fig. 9.4 on this page. As a reminder, the ground state has quantum numbers $J = 1/2$ and $F = 4$, the excited state has $J' = 3/2$ and $F' = 3$, 4, and 5, and the nuclear spin is $I = 7/2$.

1. $F = 4 \rightarrow F' = 3$: $I_r = (2(4) + 1)(2(3) + 1) \begin{Bmatrix} \frac{1}{2} & \frac{7}{2} & 4 \\ 3 & 1 & \frac{3}{2} \end{Bmatrix}^2 = \frac{7}{16} = 0.4375$

2. $F = 4 \rightarrow F' = 4$: $I_r = (2(4) + 1)(2(4) + 1) \begin{Bmatrix} \frac{1}{2} & \frac{7}{2} & 4 \\ 4 & 1 & \frac{3}{2} \end{Bmatrix}^2 = \frac{21}{16} = 1.3125$

3. $F = 4 \rightarrow F' = 5$: $I_r = (2(4) + 1)(2(4) + 1) \begin{Bmatrix} \frac{1}{2} & \frac{7}{2} & 4 \\ 5 & 1 & \frac{3}{2} \end{Bmatrix}^2 = \frac{11}{4} = 2.75$

From this math we see that the $F = 4 \rightarrow F' = 5$ is the largest transition. Using this largest transition as the reference, we can compare the size of the other two transitions to it. The $F = 4 \rightarrow F' = 4$ transition, which is the peak around 5 MHz in Fig. 9.5, is $\frac{21/16}{11/4} = \frac{21}{44} = 0.47$ times smaller than the $F = 4 \rightarrow F' = 5$ transition. Finally, the $F = 4 \rightarrow F' = 3$ transition is $\frac{7/16}{11/4} = \frac{7}{44} = 0.16$ times smaller than the $F = 4 \rightarrow F' = 5$ transition. If you look at the amplitudes of the real transitions, you'll see that these estimates are pretty close to what was measured experimentally.

While it is possible to calculate the amplitudes of the crossovers, it is much more complicated. One reason is that the number of atoms with a particular velocity, which creates the crossovers, depends on the temperature of your vapor cell. However, we at least have a fairly straightforward way of finding the relative amplitudes of the real transitions.

9.6   Example with Oxygen-16

Oxygen-16 has no nuclear spin, so there is no hyperfine structure. Shortest section ever!

9.7   Example with Oxygen-17

Oxygen-17 has a nuclear spin quantum number of $I = 5/2$, so there will be hyperfine structure. Below is a table of the seven lowest energy states for atomic oxygen.

Electron configuration	Term symbol	Energy (cm^{-1})	f (Hz)
$1s^2 2s^2 2p^4$	$^3P_2$	0	
$1s^2 2s^2 2p^4$	$^3P_1$	158.265	4.74×10^{12}
$1s^2 2s^2 2p^4$	$^3P_0$	226.977	6.80×10^{12}
$1s^2 2s^2 2p^4$	$^1D_2$	15,867.862	4.76×10^{14}
$1s^2 2s^2 2p^4$	$^1S_0$	33,792.583	1.01×10^{15}
$1s^2 2s^2 2p^3 3s$	$^5S_2$	73,768.200	2.21×10^{15}
$1s^2 2s^2 2p^3 3s$	$^3S_1$	76,794.978	2.30×10^{15}

For oxygen isotopes with no nuclear spin, such as oxygen-16, this table provides all the information about their atomic states. There is a single ground state with electronic configuration and term symbol $1s^2 2s^2 2p^4 \, ^3P_2$. However, due to hyperfine structure, the table does not tell the whole story for oxygen-17. Let's examine the ground state of oxygen-17. For this state, the total electronic angular momentum quantum number is $J = 2$ and the nuclear spin quantum number is $I = 5/2$. The possible values for the total atomic angular momentum quantum number are:

$$F = \left(2 + \frac{5}{2}\right), \ldots, \left|2 - \frac{5}{2}\right| = \frac{9}{2}, \frac{7}{2}, \frac{5}{2}, \frac{3}{2}, \frac{1}{2} \qquad (9.11)$$

We can also calculate the energy splitting using the hyperfine splitting formula. For convenience, here is the hyperfine splitting formula:

$$\Delta E = \tfrac{1}{2}KA + \frac{\tfrac{3}{2}K(K+1) - 2I(I+1)J(J+1)}{2I(2I-1)2J(2J-1)}B$$
$$K = F(F+1) - I(I+1) - J(J+1)$$
$$A = 0 \text{ unless both } I > 0 \text{ and } J > 0$$
$$B = 0 \text{ unless both } I > 1/2 \text{ and } J > 1/2. \qquad (9.12)$$

The first thing to notice is that for the $^3P_2$ ground state, both $I > 1/2$ and $J > 1/2$, so we will need both the magnetic dipole term (the term with A) and the electric quadrupole term (the term with B). For example, the hyperfine splitting for the $F = 9/2$ hyperfine level is:

$$K = F(F+1) - I(I+1) - J(J+1) = \tfrac{9}{2}\left(\tfrac{11}{2}\right) - \tfrac{5}{2}\left(\tfrac{7}{2}\right) - 2(3) = 10$$

$$\Delta E_{F=9/2} = 5A_{\text{gs}} + \tfrac{1}{4}B_{\text{gs}}$$

(9.13)

where A_{gs} and B_{gs} are the hyperfine constants for the $^3P_2$ ground state. Each energy level in the above table will have different hyperfine constants.

Next, let's explore the $^3P_0$ state, which has an energy of $226.977\,\text{cm}^{-1}$ above the ground state and a total electronic angular momentum quantum number of $J = 0$. Using the rule to find F, we find a single possible value: $F = (J+I), \ldots, |J-I| = (0+I), \ldots, |0-I| = I \rightarrow F = I = 5/2$. This only happens when a state has quantum number $J = 0$. With the constraint $F = I$, we find

$$\begin{aligned} K &= F(F+1) - I(I+1) - J(J+1) \\ &= I(I+1) - I(I+1) - 0(0+1) \\ &= 0 \\ \rightarrow \Delta E &= 0 \end{aligned}$$

(9.14)

Even though the nucleus has angular momentum, there is no energy shift when $J = 0$. This hyperfine level has the same energy as the center of gravity. However, we will still label this state with the F quantum number: $1s^2 2s^2 2p^4\,{}^3P_0\ F = \tfrac{5}{2}$. You will explore the hyperfine structure of oxygen-17 more in Problem 9.3.

Problems

9.1 Table 9.2 shows the energy shifts for the hyperfine levels with respect to the center of gravity for the $7p\,{}^2P_{3/2}$ state in cesium-133. All of the real transitions in this problem are from the $6s\,{}^2S_{1/2}\ F = 4$ ground hyperfine level.

(a) In Sect. 9.4.1, we found ΔE for the $F = 4 \rightarrow F' = 3$ real transition. Confirm ΔE for the other two real transitions.
(b) Find ΔE for the three V crossovers, similar to what we did in Sect. 9.4.2.

9.2 Below is table for four states in europium.[8] The first row represents the ground state, and the three subsequent rows are excited states. Europium has two stable isotopes, europium-151 and europium-153.

---

[8] Ground state citation: Reference [2]
[8]$P_{5/2}$ and [8]$P_{7/2}$ citation: Reference [3]
[8]$P_{9/2}$ citation: Reference [4].

Electron configuration	A_{151} (MHz)	B_{151} (MHz)	A_{153} (MHz)	B_{153} (MHz)
$4f^7 6s^2 \, ^8S_{7/2}$	-20.0523 ± 0.0002	-0.7012 ± 0.0035	-8.8532 ± 0.0002	-1.7852 ± 0.0035
$4f^7 6s6p \, ^8P_{5/2}$	-157.01 ± 0.03	74.5 ± 0.4	-69.43 ± 0.14	191 ± 2.6
$4f^7 6s6p \, ^8P_{7/2}$	-218.66 ± 0.04	-293.4 ± 0.8	-97.15 ± 0.13	-750 ± 3
$4f^7 6s6p \, ^8P_{9/2}$	-228.84 ± 0.02	226.9 ± 0.5	-101.87 ± 0.06	575.4 ± 1.5

(a) Select either the 151 or 153 isotope and one of the three excited states from the table above. Use the largest to smallest in integer steps rule to find the possible F values for the ground state and your chosen excited state.
(b) Using the hyperfine splitting equation, calculate the energy splitting for the ground state. Don't worry about the uncertainties.
(c) Using the hyperfine splitting equation, calculate the energy splitting for the excited state. Don't worry about the uncertainties.
(d) Determine the transition frequency from a single ground hyperfine level (your choice) to a single excited state hyperfine level (your choice). Report your answer in MHz. Your final answer should include f_{cog} in it.
   Note: f_{cog} is different for the two isotopes. This small difference is called an isotope shift, which we will explore in Chap. 10.

9.3 Oxygen-17 has a nuclear spin quantum number of $I = 5/2$, resulting in hyperfine structure. Sect. 9.7 contains a table that lists the seven lowest energy states.

(a) Find the possible F quantum numbers for all seven states. Hint: There are only three calculations here!
(b) For the $1s^2 2s^2 2p^4 \, ^3P_1$ state, calculate the hyperfine energy shift ΔE for each hyperfine level in terms of the hyperfine constants. Label your hyperfine constants A_{3P1} and B_{3P1}.
(c) Suppose you were to excite an oxygen atom from the $1s^2 2s^2 2p^4 \, ^3P_1$ $F = 7/2$ state to the $1s^2 2s^2 2p^3 3s \, ^3S_1$ $F' = 5/2$ state. What is the transition frequency? Your answer should look something like $f = f_{\text{cog}} + \#A_{3P1} + \#B_{3P1} + \#A_{3S1} + \#B_{3S1}$.
(d) Suppose you performed saturated absorption spectroscopy across all hyperfine levels for the $1s^2 2s^2 2p^4 \, ^3P_1 \rightarrow 1s^2 2s^2 2p^3 3s \, ^3S_1$ transition. Write the fit function for this spectrum, assuming we have no crossovers.
(e) Challenge: Include crossovers!
(f) Optional: Find the relative intensities of all the real hyperfine transitions for the spectrum in part (d). The relative intensities should be with respect to the largest amplitude transition.

9.4 Consider the $3s \, ^2S_{1/2} \rightarrow 3p \, ^2P^\circ_{3/2}$ transition in sodium-23. The little circle on the excited state term symbol is the parity of the state, a topic we don't cover

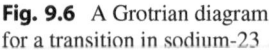

Fig. 9.6 A Grotrian diagram for a transition in sodium-23

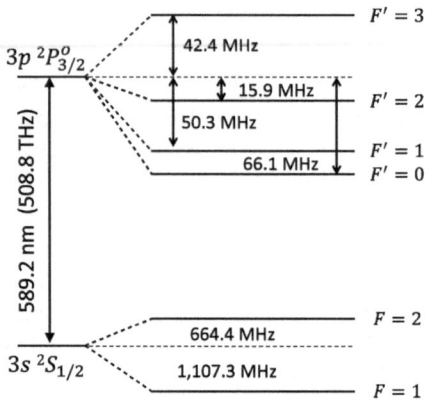

in this book. Parity is an advanced topic that is easy to say in words, but hard to understand.[9] Fig. 9.6 is a Grotrian diagram that shows the hyperfine structure of both states. We will call the hyperfine constants for the ground state A_{2S} and B_{2S} and the hyperfine constants for the excited state A_{2P} and B_{2P}.

(a) What is the nuclear spin quantum number for sodium-23?
(b) What is B_{2S}?
(c) Write the formula for the transition frequency between the 3s $^2S_{1/2}$ $F = 1$ state and the 3p $^2P_{3/2}$ $F' = 2$ state. Leave the hyperfine constants as A_{2S}, A_{2P}, and B_{2P}. You will find their values in parts (d) and (e).
(d) Determine the hyperfine constants for the 3s $^2S_{1/2}$ state.
(e) Determine the hyperfine constants for the 3p $^2P_{3/2}$ state.
(f) The center of gravity frequency is $f_{cog} = 508,848,717.1$ MHz. Determine the numerical value for the transition frequency for the 3s $^2S_{1/2}$ $F = 1 \rightarrow$ 3p $^2P_{3/2}$ $F' = 2$ transition.
(g) Optional: Find the relative amplitudes of the real transitions between the 3s $^2S_{1/2}$ $F = 2$ state and the allowed excited states.

References

1. Williams, W.D., Herd, M.T., Hawkins W.B.: Spectroscopic Study of the $7p_{1/2}$ and $7p_{3/2}$ States in Cesium-133. Laser Phys. Lett. **15**(9), 095702 (2018). https://doi.org/10.1088/1612-202X/aac97
2. Sandars P.G.H., Woodgate G.K.: Hyperfine structure in the ground state of the stable isotopes of europium. Proc. R. Soc. Lond. A. **257**, 269–276 (1960). http://doi.org/10.1098/rspa.1960.0149
3. Maruko, C., Cölmek, N., Herd, M.T., Ahrendsen, K., Cabrales, B., Cannon, G., Davis, E., Guo, X., Karani, T., Wallace, A., Wisnauckas, K., Williams, W.D.: Spectroscopic study of the $4f^7 6s^2\ ^8S_{7/2}^{\circ} - 4f^7(^8S^{\circ})\ 6s6p(^1P^{\circ})\ ^8P_{5/2,7/2}$ transitions in neutral europium-151 and europium-

[9] You should, of course, do an internet search for "parity physics" if you'd like to learn more.

153: absolute frequency and hyperfine structure. J. Opt. Soc. Am. B. **41**, 1217–1223 (2024). https://doi.org/10.1364/JOSAB.521181

4. Herd, M.T., Maruko, C., Herzog, M.M., Brand, A., Cannon, G., Duah, B., Hollin, N., Karani, T., Wallace, A., Whitmore, M., Williams, W.D.: Spectroscopic study of the $4f^7 6s^2\ ^8S^\circ_{7/2}$ − $4f^7 (^8S^\circ)6s6p(^1P^\circ)^8P_{9/2}$ transition in neutral europium-151 and europium-153: absolute frequency and hyperfine structure. J. Opt. Soc. Am. B. **39**, 2596–2602 (2022). https://doi.org/10.1364/JOSAB.467968

Isotope Shifts, Radioactive Decay, and the Nuclear Forces

10

Abstract

In this chapter, we explore the nucleus, focusing on how the number of neutrons in a nucleus influences transition frequencies and the stability of atoms. We begin by examining the effect of neutrons on transition frequencies, then shift into detailed discussions of the nuclear forces and principles governing atomic stability and radioactive decay. Various modes of radioactive decay, including neutron and proton emission, α decay, spontaneous fission, β^- decay, β^+ decay, and electron capture, are investigated, with an emphasis on the conditions under which each occurs. Additionally, we explore the nuclear shell model to understand the energetics behind different types of decays and the stability of isotopes.

Learning Goals

By the end of this chapter, you should be able to understand:

- that the number of neutrons in the nucleus can
  - cause a small shift in the spectrum,
  - make an isotope unstable.
- that unstable atoms undergo radioactive decay characterized by a time called the half-life.
- that a system always seeks to reach its lowest energy state.
- that the daughter system of a decay has lower energy than the parent system.
- that the table of isotopes is similar to the periodic table but includes all isotopes.
- that the strong nuclear force is the force that holds the nucleus together.

© The Author(s) 2025
W. Raven, *Atomic Physics for Everyone*,
https://doi.org/10.1007/978-3-031-69507-0_10

- that the weak nuclear force is the force that facilitates the annihilation and creation of particles during radioactive decay.
- the nuclear shell model and the energy arguments behind radioactive decay.

10.1   Isotope Shifts

Definitions:
- **Mass number:** The number of neutrons plus protons in an element, represented by the variable A.
- **Nucleon:** A general word for either a neutron or a proton.
- **Isotope shift:** The change in the resonance frequency of a transition between two states in an atom caused by changing the number of neutrons in the nucleus.

Different isotopes of a particular element have different numbers of neutrons. For example, europium-151 has 63 protons and 88 neutrons. The '151' in the name of the isotope indicates the total number of protons and neutrons in the nucleus: $63 + 88 = 151$. Europium-153 has 63 protons and 90 neutrons: $63 + 90 = 153$. The element name europium is just a placeholder for 'the atom with 63 protons.' To display all the information at once, we often write an isotope as follows:

$$_Z^A X_N \tag{10.1}$$

where Z is the proton number, N is the neutron number, $A = Z + N$ is the mass number, and X is the element symbol. See Appendix B for a full list of element symbols. For example, europium-153 would be written as $_{63}^{153}Eu_{90}$.

Quick Quiz
Use Appendices A or B to find the proton number for the following atoms:

1. How many neutrons does oxygen-17 have?
2. How many neutrons does hydrogen-2 have?
3. How many neutrons does beryllium-9 have?

The answers are in the footnotes.[1]

---

[1] $A = Z + N \rightarrow N = A - Z$:

1. Oxygen-17 has 8 protons, so $17 - 8 = 9$ neutrons
2. Hydrogen-2 has 1 proton, so $2 - 1 = 1$ neutron
3. Beryllium-9 has 4 protons, so $9 - 4 = 5$ neutrons

The number of neutrons in the nucleus slightly affects the energy of the states in an atom. Neutrons affect the energy of the states in two ways: they change the mass of the nucleus and slightly alter the charge distribution within the nucleus (how the protons are distributed in the nucleus). For a given isotope, the energy of each state in the atom will shift a different amount because each state has a different electron configuration and term symbol (i.e. different angular momentum and different distance from the nucleus). To study how changing the number of neutrons affects the energy of a state, we typically select the same transition and measure how the transition frequency changes as a function of neutron number. Since the energy of each state in an atom shifts a different amount, the transition frequency between those two states also changes. That shift is called the isotope shift.

Table 10.1 shows experimental results from my research group for three transitions for two different isotopes of europium. The transitions are from the $4f^7 6s^2\ ^8S_{7/2}$ ground state to the $4f^7(^8S°)6s6p(^1P°)\ ^8P_J$ states, where $J = 5/2$, $7/2$, or $9/2$. In the above electronic configurations, I did not write out the closed subshells (e.g., $1s^2 2s^2 \ldots$). Notice that adding two neutrons to the nucleus decreased the transition frequency for all three transitions. Because the two isotopes have a different number of neutrons, all of the states in europium-151 have a different energy than the states in europium-153. The difference in energy between the ground states of the two isotopes is different from the difference in energy between, say, the $J = 5/2$ excited states. Therefore, the transition frequency from the ground state to the $J = 5/2$ excited state is slightly different for the two isotopes; this is the isotope shift for this transition. Similarly, the difference in energy between the $J = 7/2$ excited states for the two isotopes is different from the difference in energy between the $J = 5/2$ excited states, so the isotope shift for this transition is different from the isotope shift for the $J = 7/2$ state. In the end, every transition in the atom will have a different isotope shift.

▶ **Important Comment** Isotope shifts are measured with respect to the center of gravity of a state. This is because each isotope for an element can have a different nuclear spin. Studying the effects of neutron number is much more convenient after removing hyperfine splitting.

Table 10.1 The isotope shifts for three transitions in the europium atom. All of the transitions are from the atomic ground state. The labels for the excited state column are the total electronic angular momentum quantum number J for the excited state: $4f^7(^8S°)6s6p(^1P°)^8P_J$. f_r is the frequency for the transition between the center of gravity of the ground state and the center of gravity for an excited state. The numbers come from references [3] and [4]

Excited state	f_r (MHz) for $^{151}_{63}Eu_{88}$	f_r (MHz) for $^{153}_{63}Eu_{90}$ (MHz)	Isotope Shift (MHz)
$J = 9/2$	652,389,757.16±0.34	652,386,593.2±0.5	3163.8±0.6
$J = 7/2$	647,708,930.6±0.6	647,705,958.4±2.6	2972.8±0.5
$J = 5/2$	642,894,493.3±0.4	642,891,693.3±0.9	2799.54±0.20

We accomplish this using Eq. 9.5. When you read studies of isotope
shifts from scientific papers, the numbers you are reading have already
accounted for hyperfine structure.

While experimentalists measure the frequency difference between isotopes, mathematically we break down the isotope shift into three expressions: the normal mass
shift $\delta f_{\text{NMS}}^{AA'}$, the specific mass shift $\delta f_{\text{SMS}}^{AA'}$, and the field shift $\delta f_{\text{FS}}^{AA'}$. The formula for
the isotope shift is:

$$\delta f^{AA'} = \delta f_{\text{NMS}}^{AA'} + \delta f_{\text{SMS}}^{AA'} + \delta f_{\text{FS}}^{AA'}, \tag{10.2}$$

where A is the mass number for one isotope and A' is the mass number of the other
isotope. Both the normal mass shift and the specific mass shift come about because
the two isotopes have different masses. The field shift comes about because the
nuclei for the two isotopes have a slightly different size. The different sizes change
the charge distribution (the distribution of protons) inside the nucleus and this leads
to a small shift in the energy levels.

The easiest way to think about normal mass shift is a classical analogy. The moon
orbiting the earth implies that the moon is orbiting about the center of the earth and
the earth is not orbiting around the moon. This is a bit of a misnomer. The moon and
the earth are actually orbiting around a point somewhere along a line between the
center of the earth and the center of the moon. Because the earth has so much more
mass than the moon, that point happens to be very close to the center of the earth.
If the moon and the earth had the same mass, they would both be orbiting about a
point directly between them. That point is called the center of mass. Analogously,
the electron is not orbiting around the center of the nucleus. Instead, the nucleus
and the electron are orbiting around the center of mass point. That point moves if
we change the mass of the nucleus. Accounting for this shift in the center of mass
point is the normal mass shift. The specific mass shift is present in atoms with more
than 1 electron. It comes about because the electrons are moving and interacting
with one another. Changing the mass of the nucleus has a small effect on those
interactions.

Out of the 3 contributions, only the normal mass shift has a simple formula:

$$\delta f_{\text{NMS}}^{AA'} = \frac{m_e}{m_p} \frac{A' - A}{AA'} f_r, \tag{10.3}$$

where A' is the mass number of an isotope (for example the 151 in europium-151),
A is the mass number of a different isotope (for example the 153 in europium-153),
$\delta f_{\text{NMS}}^{AA'}$ is the isotope shift of the isotope with mass number A' with respect to the
isotope with mass number A due to the normal mass shift, m_e is the mass of an
electron, m_p is the mass of a proton, and f_r is the transition frequency from the

center of gravity from the lower energy state to the center of gravity of the higher energy state.[2]

Example

Rubidium, an element with 37 protons, has two naturally occurring isotopes: rubidium-85 and rubidium-87.[a] There is a transition from the ground state to an excited state with a center of gravity transition frequency of 377.107 THz. What is the normal mass shift between the two isotopes?

For this example, let's make $A' = 87$ and $A = 85$:

$$\begin{aligned}
\delta f_{\text{NMS}}^{AA'} &= \frac{m_e}{m_p} \frac{A'-A}{AA'} f_r \\
&= \frac{9.11 \times 10^{-31} \text{ kg}}{1.67 \times 10^{-27} \text{ kg}} \frac{87-85}{(85)(87)} (377.107 \times 10^{12} \text{ Hz}) \\
&= 55.6 \text{ MHz}
\end{aligned} \tag{10.4}$$

Since we made $A' = 87$ and $A = 85$, a positive result means that rubidium-87 would have a larger transition frequency, at least before we account for the specific mass shift or the field shift.

---

[a]On earth, about 72% of rubidium is rubidium-85 and about 28% is rubidium-87.

The experimental value of the isotope shift between rubidium-87 and rubidium-85 for this transition is (77.583 ± 0.012) MHz, see reference [1]. We found the normal mass contribution was 55.6 MHz. The remaining shift comes from the other two contributions that are, unfortunately, difficult to calculate. However, we can try to extract information about the nucleus by measuring a large number of isotopes for a particular element.

To visually compare different isotopes, we can make a plot of the isotope shifts for a particular transition as a function of neutron number. This plot is called a King plot[3] and is a nice way to visualize how changing the number of neutrons in a nucleus affects the spectrum. Figure 10.1 shows an example of experimentally measured isotope shifts for a transition in the krypton atom that is excited using a laser near 760 nm. It was collected by Keim et al. in 1995, see reference [2]. All of

---

[2] If you are paying close attention, you might ask the question, "Does f_r refer to the transition frequency for the isotope with mass number A or A'?" The answer is neither. f_r is a calculated transition frequency assuming a nucleus with infinite mass. However, in practice, you can use f_r for either isotope or the calculated value for a nucleus with infinite mass. We can do this because isotope shifts are usually on the order of MHz to a few GHz, which is typically more than 100,000 times smaller than f_r. The formula to calculate the transition frequency assuming a nucleus with infinite mass is $\frac{m_e + M}{M} f_r$, where M is the mass of a nucleus for a particle isotope and f_r is the transition frequency for that isotope. That fraction is very close to 1.

[3] Named after the British physicist William H. King.

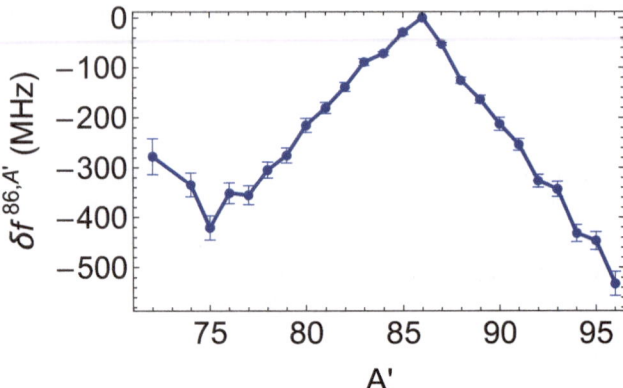

Fig. 10.1 The isotope shift of a transition in krypton near 760 nm using data from reference [2]. The isotope shifts are measured with respect to the isotope with mass number $A = 86$

the isotope shifts are measured with respect to krypton-86. Notice there is a "kink" in the graph at $A' = 86$. As physicists, we want to know why! If we truly understand the system, we should be able to calculate the isotope shifts using Eq. 10.2 and predict such a graph. The reason for the "kink" at $A' = 86$ is that nucleons fill shells in the nucleus just like electrons fill shells. We discuss nuclear shells in Sect. 10.6. There is a nuclear shell for neutrons that fills at 50 neutrons. Since krypton has 36 protons, this neutron shell fills at $A' = 86$. The "kink" at $A' = 75$ is not from a shell filling. That "kink" is thought to come from the nucleus itself beginning to deform due to lack of neutrons. There is so much one could ask and explore with this data. What questions would you ask? The wonderful thing about physics is that if we understand the system, we can interpret the data. If there is something we don't understand, this sort of data gives us a starting point to explore and learn more.

10.2   Radioactive Decay

Inside the nucleus are neutrons and protons. Experimental evidence shows that if the neutron-to-proton ratio isn't "correct," the nucleus becomes unstable and will undergo radioactive decay. Radioactive decay is often written in the form of an equation. The left hand side of the equation is the unstable isotope. We will call this the parent system. The right hand side shows all of the decay particles. We will call this the daughter system.[4] The equation looks something like this:

---

[4] Historically, parent and daughter are the names used to describe radioactive decay. The terminology also shows up in biology during cell division (parent cell and daughter cells). Sometimes the daughter system is called the decay system.

$$\text{parent} \rightarrow \text{daughter} \tag{10.5}$$

There are a number of decays that can happen:

Neutron Emission The nucleus will eject a neutron. An example of an isotope that undergoes neutron emission is helium-5 (2 protons and 3 neutrons), which decays to helium-4 (2 protons and 2 neutrons) and a neutron:

$$^{5}_{2}\text{He}_3 \rightarrow {}^{4}_{2}\text{He}_2 + \text{n}$$

In the above equation, helium-5 is called the parent nucleus and helium-4 is called the daughter nucleus. The general formula for neutron emission is:

$$^{A}_{Z}\text{X}_N \rightarrow {}^{A}_{Z}\text{X}_{N-1} + \text{n} \tag{10.6}$$

Proton Emission The nucleus will eject a proton. An example of an isotope that undergoes proton emission is scandium-39 (21 protons and 18 neutrons), which decays to calcium-38 (20 protons and 18 neutrons) and a proton:

$$^{39}_{21}\text{Sc}_{18} \rightarrow {}^{38}_{20}\text{Ca}_{18} + \text{p}$$

In the above equation, scandium-39 is called the parent nucleus and calcium-38 is called the daughter nucleus. The general formula for proton emission is:

$$^{A}_{Z}\text{X}_N \rightarrow {}^{A}_{Z-1}\text{X}_N + \text{p} \tag{10.7}$$

β^- Decay (Beta Minus Decay) A neutron will transform into a proton, an electron, and a particle called an "anti-electron neutrino" (which has the label $\bar{\nu}_e$). We haven't talked about neutrinos, but you can find more information about them with a quick internet search. For our purposes, neutrinos are very light particles with no electric charge. Both the electron and the anti-electron neutrino are ejected while the proton remains in the nucleus. The electron that is ejected from the nucleus usually has a ton of energy and is called a β^- particle. An example of an isotope that undergoes β^- decay is carbon-14:

$$^{14}_{6}\text{C}_8 \rightarrow {}^{14}_{7}\text{N}_7 + \beta^- + \bar{\nu}_e$$

In the above equation, carbon-14 (6 protons and 8 neutrons) is called the parent nucleus and nitrogen-14 (7 protons and 7 neutrons) is called the daughter nucleus. The general formula for β^- decay is:

$$^{A}_{Z}\text{X}_N \rightarrow {}^{A}_{Z+1}\text{X}_{N-1} + \beta^- + \bar{\nu}_e \tag{10.8}$$

β^+ *Decay (Beta Plus Decay or Positron Decay)* A proton will transform into a neutron, a positron,[5] and an electron neutrino (which has the label ν_e). Both the positron and the electron neutrino are ejected while the neutron remains in the nucleus. Like the electron in β^- decay, the positron has a ton of energy and is called a β^+ particle. An example of an isotope that undergoes β^+ decay is potassium-40, which is found in bananas and decays to argon-40:

$$^{40}_{19}K_{21} \rightarrow {}^{40}_{18}Ar_{22} + \beta^+ + \nu_e$$

The general formula for β^+ decay is:

$$^{A}_{Z}X_N \rightarrow {}^{A}_{Z-1}X_{N+1} + \beta^+ + \nu_e \tag{10.9}$$

Electron Capture The nucleus will steal an electron from the atom. That electron and a proton in the nucleus transform into a neutron and an electron neutrino. The electron neutrino is ejected while the neutron stays in the nucleus. An example of an isotope that undergoes electron capture is beryllium-7, which decays to lithium-7:

$$^{7}_{4}Be_3 + e \rightarrow {}^{7}_{3}Li_4 + \nu_e$$

The general formula for electron capture is:

$$^{A}_{Z}X_N + e \rightarrow {}^{A}_{Z-1}X_{N+1} + \nu_e \tag{10.10}$$

α *Decay (Alpha Decay)* Heavy nuclei can emit a cluster of two protons and two neutrons known as an α particle. This usually happens when a pair of protons is far enough away from the other protons that the Coulomb force from all the other protons is larger than the strong nuclear force (discussed in Sect. 10.4), which keeps those protons attached to the nucleus. As we will see in Sect. 10.6, an α particle is a super-stable combination of protons and neutrons. An example of an isotope that undergoes α decay is radon-222, which is a gas that leaks into some basements and decays to polonium-218:

$$^{222}_{86}Rn_{136} \rightarrow {}^{218}_{84}Po_{134} + \alpha$$

The general formula for α decay is:

$$^{A}_{Z}X_N \rightarrow {}^{A-4}_{Z-2}X_{N-2} + \alpha \tag{10.11}$$

Spontaneous Fission While most heavy nuclei undergo α decay, superheavy atoms can emit nuclei of other elements. For example, californium-252 (98 protons and 154 neutrons) can break apart into xenon-140 (54 protons and 86 neutrons),

---

[5] A positron is the antimatter partner to the electron. Antimatter is a topic in Chap. 11.

ruthenium-108 (44 protons and 64 neutrons), and 4 neutrons (notice that $140 + 108 + 4 = 252$). This is called spontaneous fission:

$$^{252}_{98}Cf_{154} \rightarrow {}^{140}_{54}Xe_{86} + {}^{108}_{44}Ru_{64} + 4n$$

In the above equation, both xenon-140 and ruthenium-108 are considered daughters.

If the balance between neutrons and protons deviates too far from a stable equilibrium, the nucleus will undergo radioactive decay. Some isotopes can undergo two or more forms of decay. For example, francium-220 undergoes α decay 99.65% of the time and β^- decay for the remaining 0.35%. Radium-226 almost always undergoes α decay, but 3.2×10^{-9}% of the time it will undergo spontaneous fission by emitting a carbon-14 nucleus. Exploring the physics behind the balance between neutrons and protons is the topic of Sect. 10.6.

Will's Rant

It is important to understand that during β^- decay, β^+ decay, and electron capture, particles cease to exist and new particles come into existence. Some books imply, for example, that a neutron is composed of a proton, an electron, and an anti-electron neutrino, as if you could look inside the neutron and find those three particles squished together. This is incorrect and drives me a little crazy, hence the rant ☺. The neutron is actually composed of three smaller particles called quarks, a topic in Chap. 11. What is important here is that during β^- decay, the neutron ceases to exist. The neutron is present one moment and gone the next. The proton, electron, and anti-electron neutrino did not exist before the decay; these particles are created during the decay. In contrast, no new particles are created or destroyed in neutron emission, proton emission, or α decay.

There is a general rule about whether or not an isotope will undergo radioactive decay. In fact, this rule is more general than radioactive decay. As a general rule of nature, things try to get to the place of lowest energy.

A General Rule of Nature

A system always seeks to reach a state of lower energy.

As an analogy, think about a ball that is sitting halfway up a bowl. When released, the ball will try to get to the lowest point in the bowl and, with friction present, will eventually settle at this lowest point. The reason the ball settled at the lowest point is because this is the place of lowest energy.

Consider the radioactive decay of carbon-14:

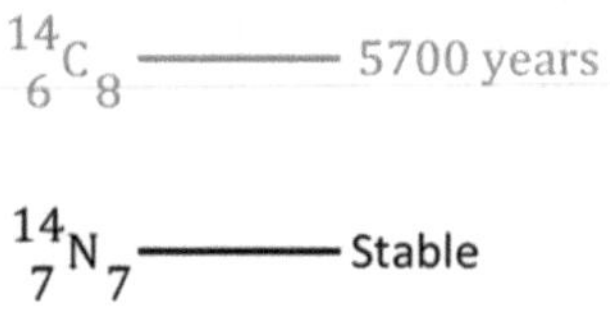

Fig. 10.2 An example of
how to think about why
carbon-14 decays to
nitrogen-14

$$^{14}_{6}C_8 \quad \underline{\hspace{2cm}} \quad 5700 \text{ years}$$

$$^{14}_{7}N_7 \underline{\hspace{1.5cm}} \text{Stable}$$

$$^{14}_{6}C_8 \rightarrow {}^{14}_{7}N_7 + \beta^- + \bar{\nu}_e$$

The daughter nucleus, nitrogen-14, has lower energy than the parent, carbon-14.
The decay happens because carbon-14 is a higher energy state than nitrogen-14,
see Fig. 10.2. The higher energy state has a lifetime of about 5700 years. After that
characteristic time, it will decay to nitrogen-14. We will discuss the mechanisms
behind radioactive decay in Sects. 10.4, 10.5 and 10.6.

All unstable isotopes have a characteristic time for decay, known as the half-life,
represented by the parameter $t_{1/2}$. Radioactive decay is a random process, but we
can state probabilities for whether an isotope has decayed. Half-life is defined as the
time it takes for there to be a 50% probability that the isotope has decayed. If the
radioactive sample is large enough, the half-life can also be thought of as the time it
takes for half of the parent nuclei to decay to the daughter nucleus. The number of
remaining parents as a function of time is given by the formula:

$$N(t) = N_0 \left(\frac{1}{2}\right)^{\frac{t}{t_{1/2}}}, \tag{10.12}$$

where N_0 is the number of radioactive atoms we start with, t is the time elapsed,
and $t_{1/2}$ is the half-life. For example, rubidium-84 has a half-life of $t_{1/2} =$
32.82 days. If we had a sample of 1,000,000,000 rubidium-84 atoms, half would
decay after 32.82 days leaving about 500,000,000 rubidium-84 atoms. Of those
remaining atoms 500,000,000 rubidium-84 atoms, half of those will decay in the
next 32.82 days. This continues until there are no radioactive atoms remaining, see
Table 10.2.

Table 10.2 How radioactive
rubidium-84 atoms decay
over time

$t_{1/2}$	Days	$N(t)$
0	0	1,000,000,000
1	32.82	500,000,000
2	65.64	250,000,000
3	98.46	125,000,000
4	131.28	62,500,000
5	164.10	31,250,000
6	196.92	15,625,000
7	229.74	7,812,500
8	262.56	3,906,250
9	295.38	1,953,125

Extra Fun

Carbon-14 is a radioactive isotope of carbon with a half-life of about 5700 years. It is created in the atmosphere when high energy neutrons from space, also known as cosmic rays, slam into the stable isotope nitrogen-14. The high energy neutron basically crashes into a proton in the nitrogen-14 nucleus, pushing the proton out while remaining behind. This radioactive carbon-14 attaches to an oxygen molecule to form radioactive carbon dioxide. Since there is a source (cosmic rays hitting nitrogen-14) and a sink (radioactive decay), the atmosphere reaches an equilibrium between the carbon-12 (the stable isotope) and carbon-14 forms of carbon dioxide. This equilibrium results in a constant isotopic abundance. If you collected a bunch of carbon dioxide from the air, about 1 molecule in every 10^{12} is radioactive. Living organisms in contact with the atmosphere breathe in, absorb, or ingest this radioactive carbon dioxide, thereby reaching equilibrium with the atmosphere. If the organism ceases interaction with the atmosphere,[6] the isotopic abundance decreases (the radioactive carbon decays without being replaced). If sometime later you dig up the formally living being and you know the starting isotopic abundance, you can quickly calculate how long that being has been removed from the atmosphere. This process is known as radiocarbon dating. There are other types of radioactive dating including radioargon dating and radiokrypton dating.

---

[6]RIP ☺.

10.3   The Table of isotopes

The table of isotopes, also known as the table of nuclides, is similar to the periodic table, but includes all known elements and isotopes. Figure 10.3 is a screenshot of the table of isotopes from a website maintained by the International Atomic Energy Agency (IAEA).[7] Figure 10.4 is a zoom in for the lighter elements. The vertical axis represents the number of protons in the isotope, while the horizontal axis represents the number of neutrons in the isotope. The entire row with 2 protons consists of helium isotopes; the entire row with 63 protons are europium isotopes. The black

---

[7] I consider the Live Chart of Nuclides website one of three essential tools for an atomic and nuclear physicist: https://www-nds.iaea.org/relnsd/vcharthtml/VChartHTML.html. The other two are the IAEA app for your phone, known as 'Isotope Browser', and the NIST spectral database, https://www.nist.gov/pml/atomic-spectra-database.

boxes indicate stable isotopes, and the colors representing the various types of radioactive decay are shown as insets in both Figs. 10.3 and 10.4. The interactive table is a lot of fun to play with.

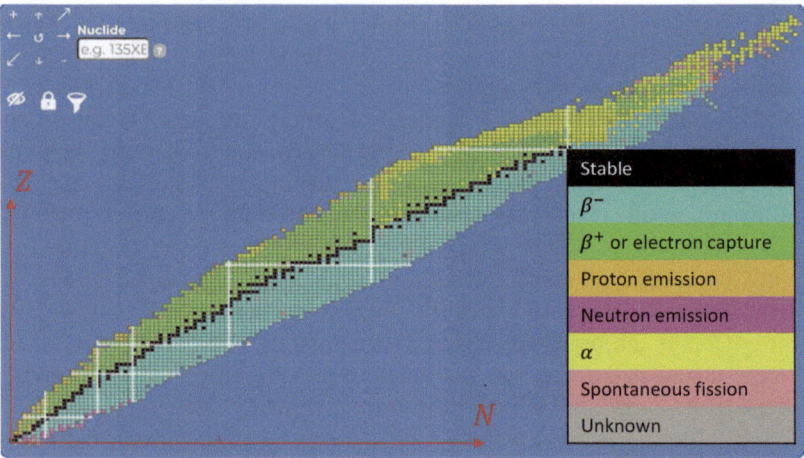

Fig. 10.3 A screenshot of the entire table of isotopes from the Live Chart of Nuclides maintained by the IAEA: https://www-nds.iaea.org/relnsd/vcharthtml/VChartHTML.html Note: I added the axes and the legend. Some isotopes have more than one type of decay. The legend shows the dominate decay

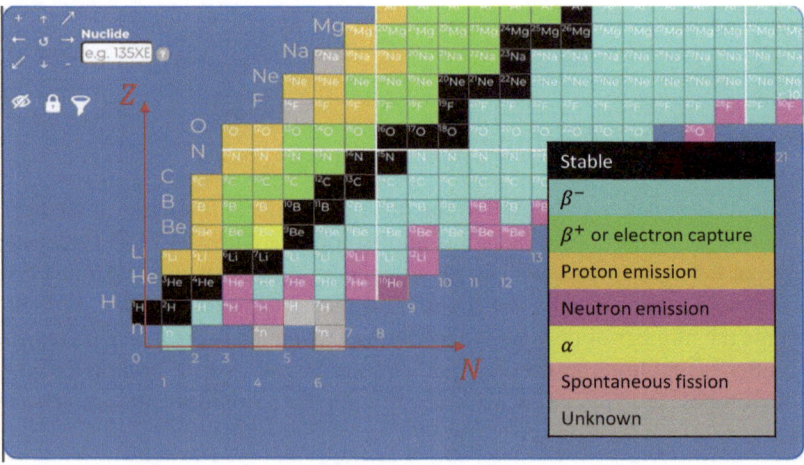

Fig. 10.4 A zoom in screenshot of the lighter elements from the table of isotopes. Note: I added the axes and the legend

10.4   The Strong Nuclear Force

> **Important Reminder**
> When two particles are interacting, there is a force between them. For
> example, an electron and a proton are attracted to each other because they have
> opposite charges. That attractive force is called the Coulomb force, which is
> also known as the electrostatic force.

We learned in Sect. 10.2 that an imbalance in the ratio of protons to neutrons
results in radioactive decay. If the balance is "good", the atom is stable. If there
is an imbalance, there will exist a daughter system that has a lower energy than
the parent system, see Fig. 10.2 for an example. To fully explore the nucleus and
atomic stability, we will first study the strong nuclear force (this section). The
strong nuclear force is an attractive force that keeps the nucleons in the nucleus. In
Sect. 10.5 we will explore the weak nuclear force, which is the force that facilitates
β^- decay, β^+ decay, and electron capture. Basically, if a process involves the
annihilation and creation of particles, the weak nuclear force is involved. Finally,
we will tie everything together in Sect. 10.6 using energy principles to explore why
some isotopes are stable and some are not.

The helium-4 nucleus has 2 protons and 2 neutrons. Since both protons have
positive charge, they repel each other due to the Coulomb force. The protons in
helium-4 are separated by about 1×10^{-15} m. This incredibly small distance is
called a femtometer, which is given the unit fm: 1×10^{-15} m $= 1$ fm. The repulsive
Coulomb force between the two protons has a magnitude of about 230 newtons
(about 50 pounds of force). Considering each proton has a mass of only $1.67 \times$
10^{-27} kg, this is huge!! If we took a free proton and acted on it with 230 Newtons
of force, the proton would accelerate at 1×10^{29} m/s^2!!!! For comparison, freefall
on earth is only about 9.8 m/s^2. The relative comparison is so enormous, I have
absolutely no idea how to convey how large that repulsive force is.

However, the protons inside a helium-4 nucleus do not fly apart.[8] Therefore,
there must be some other attractive force that is larger than the repulsive Coulomb
force to stop the protons from being pushed out of the nucleus. An intuitive guess
is that maybe gravity is keeping the protons in the nucleus. After all, gravity is the
attractive force that we interact with everyday. Remarkably, gravity is an extremely
weak force. You will explore just how weak gravity is in Problem 10.6. In addition,
neutrons have no charge, so they don't feel any Coulomb force. What is keeping
neutrons in the nucleus?

---

[8] This is a good thing!

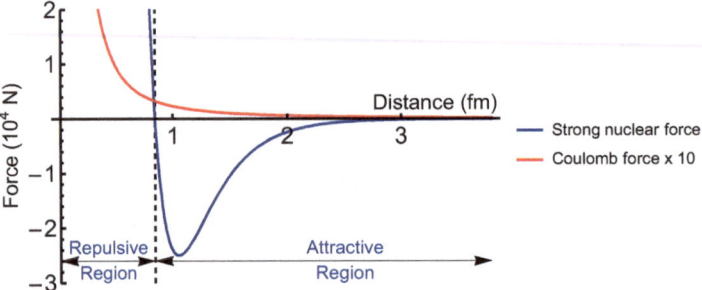

Fig. 10.5   An illustrative plot showing the forces between nucleons inside the nucleus. The strong nuclear force is attractive (negative force) above about 0.85 fm but falls off very quickly. The Coulomb force between two protons, which is increased by a factor of 10 for visibility, is always repulsive (positive force)

The force keeping the nucleons "glued" to one another is the strong nuclear force.[9] The strong nuclear force acts on both neutrons and protons, and it can be either an attractive or repulsive force. At a distance of 1 femtometer, the strong nuclear force is attractive and approximately 100 times stronger than the repulsive Coulomb force. Figure 10.5 shows a graph of both the Coulomb force and the strong nuclear force.[10] At short distances, both the strong nuclear force and the Coulomb force are repulsive. Somewhere around 0.85 fm, the nucleons start to be attracted by the strong nuclear force. The really important thing to take from this graph is that at around 1 fm, the strong nuclear force is more attractive than the Coulomb force is repulsive. The total force is the sum of the two, which for small distances, is attractive. Outside of about 5 fm, the strong nuclear force is basically 0 and the Coulomb force will dominate, pushing the protons away from each other.

Figure 10.5 is also useful for understanding why we cannot have a stable atom with more than about 200 nucleons. A proton near the edge of a large nucleus is far away from the nucleons that are on the other side of the nucleus. Because the strong nuclear force goes to zero so quickly, that proton does not feel any attractive strong nuclear force from the distant nucleons. It does, however, still feel the repulsive

---

[9] In Chap. 11, we are going to talk about quarks, which are the subparticles that make up protons and neutrons. The strong nuclear force is also what keeps the quarks glued to one another inside the nucleon. However, the strong nuclear force extends outside of a nucleon to interact with other nucleons. The force that extends out of the nucleon is often called the "strong residual force". Technically, it is still the same strong nuclear force, but there are subtle and important differences for how it behaves inside a nucleon and how it behaves between nucleons. Figure 10.5 is an illustrative plot for the strong residual force between nucleons.

[10] This strong nuclear force plot is calculated from the Reid potential, see Reference [3], which was developed and named after American physicist Roderick V Reid Jr. This is a popular model for the strong nuclear force first used in 1968.

Coulomb force from all the other protons. Consequently, this proton near the edge will be repelled away. This is why the periodic table does not have any stable elements above lead-208. The next atom on the periodic table is bismuth. Bismuth has no stable isotopes, although bismuth-209 is super long-lived with a half-life of 2×10^{19} years.

An Analogy
The Coulomb force is a long-range force. Two protons will weakly repel each other even if they are 10 meters apart. The strong nuclear force is similar to a contact force such as two pieces of Velcro. When the two pieces of Velcro are touching, they stick together really well. The moment they are no longer touching, the force between the two is zero.

So, why do we need neutrons for a stable nucleus? Neutrons add additional attractive strong nuclear force to the nucleus without adding any additional repulsive Coulomb forces. A nucleus with only 2 protons, which would be helium-2, is extremely unstable. The Coulomb force, while small compared to the strong nuclear force, is still large enough to push the two protons apart. So, the decay of helium 2 would be two protons flying away in different directions. Adding a neutron to the nucleus adds additional strong nuclear force (attractive) without adding any additional repulsive Coulomb forces. Both helium-3 (2 protons and 1 neutron) and helium-4 (2 protons and 2 neutrons) are stable. Helium-5 (2 protons and 3 neutrons) is unstable. But why is helium-5 unstable? An extra neutron should just add additional attractive strong nuclear force with no additional repulsive Coulomb force. In the end, it comes down to the energy analogy: there is a daughter system with lower energy than helium-5. We will explore this more in Sect. 10.6.

Strong Nuclear Force Summary
Inside the nucleus, the strong nuclear force is the mechanism that keeps nucleons in the nucleus. It is a short-range force. Neutrons add additional attractive strong nuclear force without adding any additional repulsive Coulomb force. If there are too many protons compared to neutrons or the nucleus is too large, proton emission or α decay may occur.

Fun Fact

If the strong nuclear force were only 2% larger, helium-2 could undergo β^- decay to hydrogen-2 (also known as deuterium or heavy hydrogen) instead of breaking apart into 2 protons.[11]

---

[11]See reference [4].

10.5   The Weak Nuclear Force

The strong nuclear force holds nucleons in the nucleus. If the balance between protons and neutrons is not correct, the nucleus will be unstable and undergo radioactive decay (more on this in Sect. 10.6). We have hinted at possible mechanisms behind proton emission and α decay (the Coulomb force overwhelms the strong nuclear force), but nothing we have talked about so far explains the forces behind the annihilation or creation of particles (i.e., β^-, β^+, and electron capture; see the beginning of Sect. 10.2 for a recap of the types of radioactive decays). To explore those decays, we need to understand the weak nuclear force.

Important

If the radioactive decay process involves the annihilation and creation of particles, the weak nuclear force is involved.

Let's start by reviewing a few important concepts. The Coulomb force acts only on charged particles. It is a long-range force and can be either an attractive force (opposite charges) or a repulsive force (same charges). The strong nuclear force acts on protons and neutrons but not on electrons. It is a short-range force with a range on the order of a few femtometers and is, for our purposes, an attractive force.

The weak nuclear force, sometimes referred to simply as "the weak force," acts on electrons, protons, and neutrons. Like the Coulomb force, it can be either attractive or repulsive. However, unlike the Coulomb force, it is an *extremely* short-range force. While the strong nuclear force has a range on the order of a few femtometers, the weak nuclear force has a range on the order of 1×10^{-18} m, or about 0.1% of the diameter of a proton. This distance is called an attometer (am): 1×10^{-18} m $= 1$ am.

As you can probably guess from the name, the weak nuclear force is weak compared to the strong nuclear force and the Coulomb force. Consider two protons about 1 fm apart. The weak nuclear force is about 1 million times weaker than the strong nuclear force. For stable atoms, the strong nuclear force dominates the weak nuclear force, but for atoms with a few too many neutrons or protons, the weak nuclear force can play an important role.

To explore the importance of the weak nuclear force, let's discuss the neutron. A neutron outside of a nucleus, called a free neutron, will undergo radioactive decay with a half-life of about 10.2 minutes according to the equation:

$$n \rightarrow p + \beta^- + \bar{\nu}_e, \tag{10.13}$$

where n is the neutron that decays into a proton p, a β^- particle (a high energy electron), and an anti-electron neutrino $\bar{\nu}_e$.

A Reminder of an Important Rule and an Analogy from Sect. 10.2

A General Rule of Nature A system always seeks to reach a state of lower energy.

The Analogy Think about a ball that is sitting half way up a bowl. When released, the ball will try to get to the lowest point in the bowl and, with friction present, will eventually settle at this lowest point. The reason the ball settled at the lowest point is because this is the place of lowest energy. Both gravity and friction were the mechanisms (forces) that got the ball to settle at the bottom of the bowl. Gravity pulled the ball down, and friction dissipated the extra energy into heat. The ball went from a higher energy state to a lower energy state.

The proton, β^- particle, and anti-electron neutrino (i.e., the daughter system) have lower energy than the free neutron (the parent system). Something had to facilitate this system going from higher energy to lower energy. The force behind this interaction is the weak nuclear force. So, according to the general rule of nature, a free neutron "wants" to get to a place of lower energy, and the weak nuclear force is the force that helps it get there.

A free neutron will undergo radioactive decay facilitated by the weak nuclear force. For a neutron inside a nucleus, there is also the strong nuclear force from other nucleons interacting with the neutron. The question to ask is, "Is there a place of lower energy for the system to go?" If the answer is yes, the weak nuclear force will eventually get the system to a lower energy. If the answer is no, then the weak force has no place to push the system. So, the weak nuclear force will cause the free neutron to cease existing and create a proton, β^- particle, and anti-electron neutrino. In contrast, there is no place of lower energy for, for example, a helium-4 nucleus. So, the helium-4 nucleus is stable.

If a neutron in a helium-4 nucleus were to undergo β^- decay, the equation would be:

$$\ce{^4_2He_2} \rightarrow \ce{^4_3Li_1} + \beta^- + \bar{\nu}_e$$
This does not happen!

The system on the right of this equation (the lithium-4 atom, the β^- particle, and the anti-electron neutrino) has a higher energy than the helium-4 atom. Therefore, this process does not occur.

Current Summary
As a general rule of thumb:

- If there are way too many neutrons in the nucleus, a neutron will escape (more on this in Sect. 10.6) to get the system to a point of lower energy.
- If there are a few too many neutrons in the nucleus, the weak nuclear force will facilitate β^- decay to get the system to a point of lower energy.
- If there are way too many protons in the nucleus, the Coulomb force will push a proton out of the nucleus to get the system to a point of lower energy.

A free proton is stable[9] because a free proton is already the system of lowest energy. However, a proton can decay inside the nucleus. How?!? The answer always goes back to the energy argument: a system wants to get to the place of lowest energy. If a nucleus has a few too many protons, the weak nuclear force will find a way to that place of lower energy. In this case, a proton can transform into a neutron, a positron (more on positrons in Chap. 11), and an electron neutrino. This is called β^+ decay. As an example, consider the following decay for oxygen-15:

$$\ce{^{15}_8O_7} \rightarrow \ce{^{15}_7N_8} + \beta^+ + \nu_e \tag{10.14}$$

This decay, which has a half-life of about 2 minutes, is used in Positron Emission Tomography (PET) scans to measure blood flow and oxygen metabolism.[10] A proton in the oxygen-15 nucleus transforms into a neutron, a β^+ particle (positron), and an electron neutrino. The right-hand side of that equation is lower energy than the left-hand side, so the weak nuclear force will make this decay happen.

The daughter particle, nitrogen-15, is the lowest energy state for this system, so it does not undergo any further type of decay. Therefore, neither of the following occurs:

---

[9] As far as we know.

[10] Due to its short half-life, it is made in an accelerator on site (usually in the basement of the hospital) and brought immediately to the patient.

$$^{15}_{7}N_8 \rightarrow {}^{15}_{6}C_9 + \beta^+ + \nu_e$$
$$^{15}_{7}N_8 \rightarrow {}^{15}_{8}O_7 + \beta^- + \bar{\nu}_e$$

Neither of these happen!

Both of these processes have daughter systems with higher energy than the parent system, so they will never occur.

There is one more process in which the weak nuclear force plays an important role. For some atoms, the electrons "orbiting" the nucleus can come too close. If that happens **and** there is a daughter system with a lower energy, that electron and a proton in the nucleus will transform into a neutron and an electron neutrino. This is called electron capture. The weak nuclear force, once again, is the force that is facilitating this process. I want to emphasize that the proton and electron do not combine to form a neutron. The proton and electron literally stop existing, and a neutron and an electron neutrino start existing (see Will's rant in Sect. 10.2). An atom that undergoes electron capture decay is the beryllium-7 atom, which has a half-life of about 53 days. The equation to describe beryllium-7 decay is:

$$^{7}_{4}Be_3 + e \rightarrow {}^{7}_{3}Li_4 + \nu_e, \tag{10.15}$$

where $^{7}_{3}Li_4$ is the daughter lithium-7 atom. Once again, this process can only occur because the daughter system has lower energy than the parent system.

Summary for the Weak Nuclear Force
If particles are going to be annihilated and created to get the system to a place of lower energy, the weak nuclear force is the force that facilitates the process.

Current Summary
As a general rule of thumb:

- If there are way too many neutrons in the nucleus, a neutron will escape (more on this in Sect. 10.6) to get the system to a point of lower energy.
- If there are a few too many neutrons in the nucleus, the weak nuclear force will facilitate β^- decay to get the system to a point of lower energy.
- If there are a few too many protons in the nucleus, the weak nuclear force will facilitate either β^+ decay or electron capture to get the system to a point of lower energy.
- If there are way too many protons in the nucleus, the Coulomb force will push a proton out of the nucleus to get the system to a point of lower energy.

10.6   The Nuclear Shell Model and Energetics

When we talked about the strong nuclear force, we had an unanswered question: "Why is a nucleus with too many neutrons unstable?" After all, neutrons only add an attractive force to the system. There was an explanation about the existence of a lower energy state, but that explanation, at least to me, felt unsupported. In this section, we will explore what happens when more and more nucleons are added to a nucleus. We will also see why, from an energy viewpoint, different types of decays occur and why stable lighter atoms have about equal numbers of protons and neutrons while stable heavier atoms have more neutrons than protons, see the black boxes in Fig. 10.3. To do so, we will rely on the general rule of nature and the Pauli Exclusion Principle, see Chap. 8, Sect. 8.5 for a review of the Pauli Exclusion Principle.

> **Important Reminder**
>
> **Pauli Exclusion Principle** Two fermions cannot simultaneously occupy the same quantum state; that is, no two fermions can have the same set of quantum numbers within a quantum system. Electrons, protons, and neutrons are all fermions with spin quantum number 1/2.

In Chap. 8, we learned that electrons fill shells. The 1s subshell can hold two electrons. Using the ket notation $|n \ \ell \ m_\ell \ s \ m_s\rangle$, the two electrons have quantum numbers $|1 \ 0 \ 0 \ \frac{1}{2} \ \frac{1}{2}\rangle$ and $|1 \ 0 \ 0 \ \frac{1}{2} \ \text{-}\frac{1}{2}\rangle$. Notice they have different quantum numbers as required by the Pauli Exclusion Principle. All six electrons in a filled 2p subshell also have different quantum numbers. The 1s electrons are in a lower energy state than the 2p electrons, but the Pauli Exclusion Principle prevents any more electrons from being in the 1s subshell. The dominant interaction between the electrons and the nucleus is the electromagnetic force, so this force determines how many

Table 10.3 Electron configuration for shells and subshells

Shell	Subshell	Subshell max electrons	Shell max electrons
1	1s	2	2
2	2s	2	8
	2p	6	
3	3s	2	18
	3p	6	
	3d	10	
4	4s	2	32
	4p	6	
	4d	10	
	4f	14	

subshells there are for a particular shell. Table 10.3 shows the maximum number of electrons that can fit into each subshell and shell. I want to emphasize that every electron has a unique set of quantum numbers. Neon has a total of ten electrons to completely fill the $n = 1$ and $n = 2$ shells. Each of those ten electrons has a unique set of quantum numbers. As a reminder, electrons do not have to completely fill a shell before starting to fill the next shell (see Fig. 8.3). For example, potassium has a single electron in the 4s subshell with no electrons in the 3d subshell. This ordering is determined by the electromagnetic force.

Protons and neutrons are also fermions with a spin quantum number of $1/2$. Therefore, they also fill nuclear shells and subshells. Since protons and neutrons are different particles, they fill individual nuclear shells; protons fill the proton shells and neutrons fill the neutron shells. Isotopes with filled nuclear shells and subshells tend to be more stable. This stability occurs at so-called "magic numbers." The magic numbers for protons indicate how many protons are in the nucleus when this extra stability occurs. The magic numbers for protons are 2, 8, 20, 28, 50, 82, and 114. The magic numbers for neutrons are 2, 8, 20, 28, 50, 82, and 126. Oxygen-16 (8 protons and 8 neutrons) is called doubly magic because it has both a filled proton shell and a filled neutron shell. Oxygen-16 is an extra stable nucleus.

The configuration of the nuclear shells is very different from the electron shells because the nuclear shells are determined by both the strong nuclear force and the electromagnetic force. However, every proton (or neutron) in the nucleus has a unique set of quantum numbers and fills the proton (or neutron) shells. The last magic number for protons (114) is smaller than the last magic number for neutrons (126). This is because protons have an additional Coulomb force acting on them, and that additional force causes a different structure for the shells. I want to emphasize that these are not electron shells! Protons and neutrons do not fill up their nuclear shells according to the Madelung rule (see Fig. 8.3).

Let's use this as a starting point to understand nuclear stability. Figure 10.6 shows a planetary model for protons and neutrons filling their respective shells. As a reminder, the planetary model is not correct. Protons and neutrons are quantum mechanical particles that have wave-like properties. Nonetheless, this nuclear shell model can help us understand the stability of the nucleus.

The number of nucleons that can occupy a subshell is indicated on the left. The lowest energy proton shell can hold two protons (one spin up and one spin down) and the lowest energy neutron shell can hold two neutrons. This is the first magic number, 2. The next shell can hold a total of six nucleons distributed among two subshells. The sum of all protons (or neutrons for the neutron shells) that completely fill the first two shells is the second magic number, $2 + 6 = 8$. The third grouping of subshells holds a total of twelve nucleons, a magic number of $2 + 6 + 12 = 20$, and so on. Notice that as the energy increases, the neutron shells are lower in energy than the proton shells. This is because the Coulomb force acts on the protons, but not on the neutrons, and also because neutrons have a slightly larger mass than protons. It should be noted that the energy spacings are not correct; they are greatly simplified for us to explore concepts. There are also higher energy shells that are not shown in these diagrams.

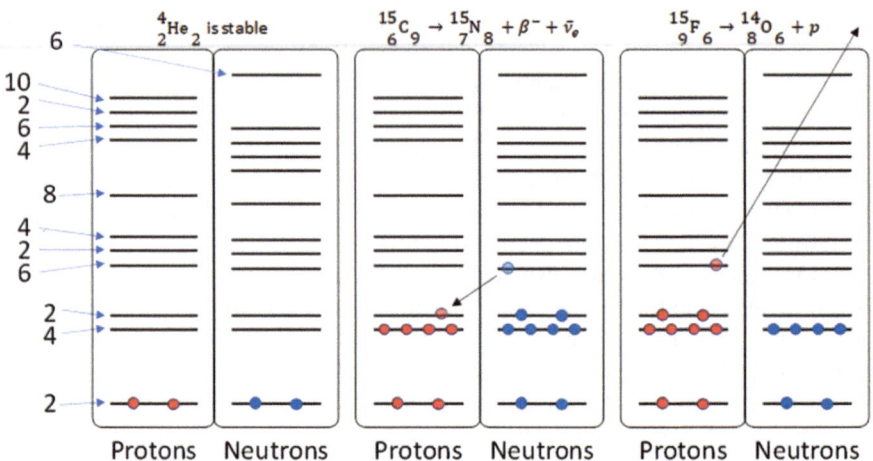

Fig. 10.6 The nuclear shells for protons and neutrons. Starting from the left, helium-4 is a stable isotope. The system is in its lowest energy state. The middle diagram is an example of $\beta-$ decay. The right diagram is an example of proton emission

Starting from the left diagram in Fig. 10.6, we see that helium-4 is a stable isotope. In fact, this is a super-stable, doubly magic nucleus because it perfectly fills the lowest nuclear shell for both the protons and the neutrons. There isn't any place of lower energy for any of the nucleons. The helium-4 nucleus is also the α particle from α decay. The middle diagram is an example of β^- decay. The system has a place of lower energy if the highest-energy neutron transforms into a proton (plus a β^- particle and anti-electron neutrino). One could ask the question, "Why is this β^- decay and not neutron emission?" The answer always come back to energy. For this scenario, it is energetically more favorable for the weak nuclear force to facilitate β^- decay than the neutron to just leave the nucleus. The right diagram is an example of proton emission. The highest-energy proton has a place to go to bring the system to a point of lower energy. It can leave the nucleus or undergo β^+ decay. In this scenario, the Coulomb force makes proton emission energetically more favorable than β^+ decay, so that is what happens. Finally, the left diagram in Fig. 10.7 shows an example of β^+ decay. In this scenario, it is energetically more favorable for the weak nuclear force to facilitate the decay of scandium-41 to calcium-41. And this is the essence of stable versus unstable nuclei. If there is a state of lower energy, the system will try to get there. Sometimes the mechanism is proton emission, neutron emission, α decay, or spontaneous fission. Other times, the weak nuclear force facilitates the annihilation and creation of particles. In the end, if there is a state of lower energy, the nucleus will decay to this lower state.

There is another interesting concept I'd like to explore with you. For smaller mass, stable nuclei, there are roughly equal numbers of neutrons and protons. This can be seen by the black boxes (stable nuclei) in Fig. 10.3, which make a linear line with a slope of one for smaller mass nuclei. This is because, for smaller masses, the nuclear energy levels for the protons and neutrons are about the same. This

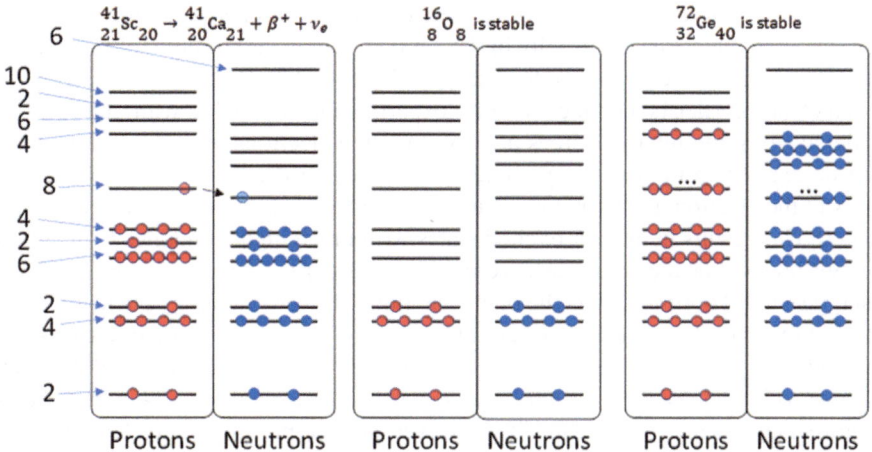

Fig. 10.7 Starting from the left, scandium-41 is an example of β^+ decay. The middle diagram and right diagram are examples of stable nuclei. Oxygen-16 is stable and has 8 protons and 8 neutrons. Germanium-72 is also stable, but it needs 40 neutrons to balance the energy of the 32 protons. The three dots on the subshell that can hold 8 nucleons are indicating the shell is filled; I couldn't fit 8 nucleons on the line

is shown in the left diagram in Fig. 10.6 for helium-4 and the middle diagram in Fig. 10.7 for oxygen-16. However, as more protons are packed into the nucleus, the Coulomb force becomes larger and larger, increasing the nuclear energy levels for the protons. As such, more neutrons are needed to balance the energy of the protons and neutrons. This is an equivalent way of saying more neutrons are needed to provide a larger strong nuclear force to overcome this increase in the Coulomb force repulsion. An example of a heavier nucleus is shown in the right diagram of Fig. 10.7. Germanium-72 needs eight more neutrons than protons to create a stable balance between the total proton energy and the total neutron energy. In fact, germanium-70, -72, -73, and -74 are all stable.[11] Notice how the diagrams for all the stable nuclei have approximately equal energy for the highest-energy neutrons and protons while the unstable nuclei do not.

If the nucleus gets too big, there is always a state of lower energy. The largest mass nucleus that is stable is lead-208. Lead-208 has a proton number of 82 and a neutron number of 126, which are both magic numbers. There are also isotopes that have two or more decay paths to states of lower energy. This is more likely for larger mass nuclei. For example, bismuth-212 can undergo both α decay (36% of the time) and β^- decay (64% of the time). For the heaviest of atoms, the dominant decay is not the annihilation and creation of particles, but particles simply leaving the nucleus to reach a lower energy state, see Fig. 10.3 or the online table of isotopes.

The nucleus is an amazing and wonderful system to study. It is such a rich system to explore and is needed to understand the world of the super small. Atomic

---

[11] Germanium-71 decays via electron capture, which is facilitated by the weak nuclear force.

physicists tend to use lasers to excite electrons to higher energy atomic states. We care about the nucleus, but usually in the context of how the structure of the nucleus affects electronic energy levels, for example, isotope shifts discussed in Sect. 10.1.

Nuclear physicists conduct similar experiments on nuclei with the goal of understanding the internal forces at play. Just like electrons, protons and neutrons can be excited to higher energy states. Compared to the Coulomb force and the mass of the electron, the strong nuclear force and the mass of the nucleons are much larger! This results in nuclear excited states with much higher energies than atomic energy levels. For the most part, we can't use lasers to excite nucleons to higher energy states. Thorium-229 is the one exception. The lowest energy nuclear excited state of thorium-229 requires a laser with a wavelength $\lambda = 150$ nm, or a frequency of 2.00×10^{15} Hz. That laser isn't easy to make or use, but it is possible! The next isotope with the lowest lying nuclear excited state is protactinium-234, which would require a laser with $\lambda = 16.8$ nm ☹. However, nuclear physicists are clever and use other methods to study nuclear excited states. For example, nuclear physicists can use accelerators and high-energy collisions to excite a nucleus and then measure the energy of the high-energy photons that are emitted. They also use radioactive decay. For this technique, some parent systems decay to excited nuclear levels that then decay via high-energy photons. If you find the nucleus as fascinating as I do, nuclear physics might be the field for you!

Problems

10.1 Consider the transition in the beryllium atom $1s^2s^2\ {}^1S_0 \rightarrow 1s^2s2p\ {}^1P_1$. The resonance frequency for this transition is $f_r = 1,276,080,100$ MHz.

(a) Calculate the normal mass shift as a function of mass number for $A' = 7$ to $A' = 12$ with respect to the stable isotope beryllium-9.
(b) Make a (modified) King plot of the normal mass shift versus A'.

10.2 Write the decay equation for the following radioactive decays:

- β^- decay:
  - Potassium-40 (${}^{40}_{19}K_{21}$): Potassium-40 is the largest source of natural radioactivity in animals, including humans. About 89% of all potassium-40 decay is β^- decay.
  - Rubidium-87 (${}^{87}_{37}Rb_{50}$): Rubidium-87 is used in rubidium-strontium dating, a radiometric dating technique used to determine the age of rocks and minerals.
- β^+ decay:
  - Sodium-22 (${}^{22}_{11}Na_{11}$): Sodium-22 is used as a calibration source for positron emission tomography (PET) scans.
  - Carbon-11 (${}^{11}_{6}C_5$): Carbon-11 is used in PET scans to detect sites of prostate cancer.

- Electron capture:
  - Potassium-40 ($_{19}^{40}K_{21}$): Potassium-40 is used in potassium-argon dating. About 11% of all potassium-40 decay is β^- decay.
  - Beryllium-7 ($_4^7Be_3$): Beryllium-7 is used in cosmogenic isotope studies to understand solar activity and atmospheric processes. It is also the lightest element to undergo electron capture.
- Alpha decay:
  - Uranium-238 ($_{92}^{238}U_{146}$): Uranium-238 produces about 40% of the radioactive heat produced in the earth.
  - Thorium-232 ($_{90}^{232}Th_{142}$): Thorium-232 is used in thorium reactors and in dating geological formations through thorium-lead dating.

10.3 (Carbon Dating) The current atmospheric isotopic abundance is $\frac{\text{carbon-14}}{\text{carbon-12}} = 10^{-12}$.

(a) Using the table of isotopes or online resources, find the half-life of carbon-14.

(b) A mummy was recently found in what was the ancient city of Memphis. The egyptologist who led the expedition sends a sample of the mummy to a radiocarbon dating specialist for carbon dating. The results came back as isotopic abundance, and not a date. The isotopic abundance of the sample is $\frac{\text{carbon-14}}{\text{carbon=12}} = 5.8 \times 10^{-13}$. Assuming the atmospheric isotopic abundance stays constant in time,[12] estimate how long ago the individual who is now a mummy died?

10.4 Locate rubidium-84 in the table of isotopes. Given a sample of 1 billion rubidium-84 atoms, determine the composition of atoms remaining after 100 days. Hint: Rubidium-84 decays into both krypton-84 and strontium-84.

10.5 A decay chain is a series of radioactive decays that shows the sequential process of decays. It is also known as a "radioactive cascade." For example, oxygen-20 undergoes β^- decay to fluorine-20, which then undergoes β^- decay to neon-20, which is stable. So, the decay chain is:

$$^{20}_{8}O_{12} \xrightarrow{\beta^-} {}^{20}_{9}F_{11} \xrightarrow{\beta^-} {}^{20}_{10}Ne_{10}$$

Construct a decay chain for thorium-232, considering only the dominant decay mode at each step. When reaching bismuth-212, follow the decay path to thallium-208. The decay chain should terminate at lead-208.

---

[12] It is not constant, but we have ways to calibrate how the ratio has changed in times as well as geographical variations.

10.6 Newton's law of universal gravitation is a model of the gravitational force between two objects with mass. Mathematically, the force between two objects with mass m_1 and m_2 is:

$$F_G = \frac{Gm_1m_2}{r^2}, \tag{10.16}$$

where $G = 6.674 \times 10^{-11} \frac{\text{N m}^2}{\text{kg}^2}$ is a constant of nature, known as the gravitational constant, and r is the distance between the two objects. The gravitational force between two objects is always attractive.

Coulomb's law is a model of the force between two objects with charge. Mathematically, the force between two objects with charge q_1 and q_2 is:

$$F_C = \frac{kq_1q_2}{r^2}, \tag{10.17}$$

where $k = 8.988 \times 10^9 \frac{\text{N m}^2}{\text{C}^2}$ is a constant of nature, known as the Coulomb constant, r is the distance between the two objects, and the unit C stands for Coulomb. A proton has a charge of $+1.602 \times 10^{-19}$ C and an electron has the same magnitude charge but opposite sign, -1.602×10^{-19} C.

(a) Consider two protons in a nucleus. How much larger is the Coulomb force compared to the gravitational force?
Hint: Take the ratio of the two forces first. The distance between the two protons will cancel out.

(b) For an object with mass m_1 at the surface of the earth, Newton's law of universal gravitation is:

$$F_G = \frac{Gm_1m_E}{R_E^2}, \tag{10.18}$$

where $m_E = 5.972 \times 10^{24}$ kg is the mass of the earth, $R_E = 6.378 \times 10^6$ m is the radius of the earth. This formula can be simplified to:

$$F_G = m_1 g, \tag{10.19}$$

where $g = \frac{Gm_E}{R_E^2}$. What is the numerical value of g? The units should simplify to m/s^2.
Hint: A newton N is the same thing as $\frac{\text{kg m}}{\text{s}^2}$.

(c) Calculate the force of gravity between you and the earth. The conversion between pounds and kilograms is 1 kg = 2.205 lbs.

(d) Find the ratio of the Coulomb force of two protons separated by 1 fm to your answer in part (c). For context, you are much, much more massive than a proton!
(e) The magnitude of the strong nuclear force between two nucleons separated by 1 fm is about 24,000 N. Find the ratio of this strong nuclear force compared to your answer in part (c).

10.7 (Advanced Math Problem: Connecting Half-Life to Lifetime) Equation 10.12 is the formula most everyone uses for radioactive decay. However, it can be useful to convert this equation to one of exponential decay: $N(t) = N_0 e^{-t/\tau}$, where τ is the lifetime. This is the same form we used to model the probability that an electron in an excited state decays to a lower energy state, see Sect. 3.3!

(a) Carbon-14 has a half-life of 5700 years. What is the lifetime?
(b) Make a graph of both equations. If your answer to part (a) is correct, the two functions should graph identically.

References

1. Barwood, G.P., Gill, P., Rowley, W.R.C.: Frequency measurements on optically narrowed Rb-stabilised laser diodes at 780 nm and 795 nm. Appl. Phys. B **53**, 142–147 (1991). https://doi.org/10.1007/BF00330229
2. Keim, M., Arnold, E., Borchers, W., Georg, U., Klein, A., Neugart, R., Vermeeren, L., Silverans, R.E., Lievens, P.: Laser-spectroscopy measurements of 72–96Kr spins, moments and charge radii. Nucl. Phys. A **586**(2), 219–239 (1995). https://doi.org/10.1016/0375-9474(94)00786-M
3. Reid, R.V.: Local phenomenological nucleon-nucleon potentials. Ann. Phys. **50**(3), 411–448 (1968). https://doi.org/10.1016/0003-4916(68)90126-7
4. Bradford, R.A.W.: The effect of hypothetical diproton stability on the universe. J. Astrophys. Astron. **30**, 119–131 (2009). https://doi.org/10.1007/s12036-009-0005-x

The Standard Model of Particle Physics

11

Abstract

In this chapter, we explore the fascinating world of particle physics and the Standard Model of Particle Physics. We discuss the constraints of the Schrödinger equation and the necessity of quantum field theory, introducing key concepts like antimatter, vacuum fluctuations, and Feynman diagrams. The chapter details the fundamental particles and forces in the Standard Model and highlights unresolved questions, such as dark matter, dark energy, and the integration of gravity. Additionally, we examine the role of virtual particles and the impact of vacuum fluctuations on our understanding of particle interactions. We aim to provide a comprehensive overview of the current state of particle physics and the exciting challenges that lie ahead.

Learning Goals

By the end of this chapter, you should be able to understand:

- that quantum mechanics does a great job of describing the world of the super small, but it is not perfect.
- that the Standard Model of Particle Physics is a conceptual model of the world of the super small; the mathematical framework for it is called Quantum Field Theory.
- that the Standard Model of Particle Physics is not complete. There are problems with it, but we aren't entirely sure how to make a more complete model. There are lots of ideas, but so far, no one has been able to experimentally confirm a better model.

© The Author(s) 2025
W. Raven, *Atomic Physics for Everyone*,
https://doi.org/10.1007/978-3-031-69507-0_11

11.1   Problems with Quantum Mechanics

We spent a lot of this book conceptually thinking about atoms. Many of the conceptual ideas come from quantum mechanics, which is mathematically described by the Schrödinger equation. Whenever you work with a model, it is important to know the limitations of your model. For example, you can't use classical physics, like Newton's second law, to describe the world of the super small. For things that are moving very fast,[1] we need Einstein's theory of special relativity. The goal of fundamental science is to develop a model that accurately describes nature. Since we now have a good conceptual background on the world of the super small, let's discuss the limitations of the quantum mechanical model.

- **Things moving fast:** The Schrödinger equation is only for slow-moving-particles. However, physicists have incorporated the ideas and mathematical models of special relativity into the Schrödinger equation. But none of these did a complete job. If you are interested in this topic, you should search the internet for "the Dirac equation" and "Klein–Gordon equation." Just a heads-up, the Wikipedia pages for these two equations are extremely mathematical. For completeness, folks also use something called perturbation theory that helps approximate the effects of fast-moving particles. It can do a really good job, but, in the end, perturbation theory is just an approximation.
- **Radioactive decay:** The mathematical model behind quantum mechanics, the Schrödinger equation, conserves particle number.[2] This will be one of the first things you prove if you take a quantum mechanics class.[3] Particle conservation means if a particle exists at one point in time, it has always existed and will always exist. We know this is not true. We create and destroy particles all the time! This is the whole purpose of a particle accelerator. In addition, we already talked about radioactive decay. For example, during β^- decay, a neutron stops existing and a proton, electron, and anti-electron neutrino come into existence. That means we went from 1 neutron to 0 neutrons. Particle number is definitely not conserved.
- **Excited state decay:** An atom that has an electron in an excited state will decay to a lower energy state and emit a photon. This process is random and mathematically modeled by an exponential decay (this was talked about back in Chap. 3 if you need a refresher). However, to get an atom to decay, it needs to be acted upon from the outside. Think about how we need force (or torque) to change an object's momentum (or angular momentum). But an excited state atom seems to decay all by itself with no external interactions.
- **Quarks:** Physicists discovered that a proton is composed of three particles known as quarks. While we will talk more about quarks in Sect. 11.5, the really

---

[1] By very fast, we mean something like 10% of the speed of light or faster.

[2] This is also called conservation of probability.

[3] At least you should. It is super important to always know the limitations of your model.

important thing here is that the mass of the three individual quarks is only about 1% of the mass of the proton. Think about that for a minute. Imagine I hand you 3 steel balls. You weigh each one individually and determine two of the balls have a mass of 2.2 grams and one of the balls has a mass of 4.7 grams. Now you put all 3 steel balls on the scale. A logical guess would say the total mass should be 2.2 grams + 2.2 grams + 4.7 grams = 9.1 grams. Instead, you find that the mass is 938 grams. Replace grams with a much smaller unit for mass and you have what actually happens ...

- **The speed of causality:** Imagine you had two charged particles that are separated by a few meters. Next you nudge one of the charged particles. According to the Schrödinger equation (and also classical electromagnetic theory), the other particle reacts instantaneously. You could repeat this experiment with one of the charged particles on earth and the other on Mars and get the same result. The two charged particles feel a force from the other charged particle and if you move one particle, the other will instantaneously feel a force that pushes or pulls on it. This violates special relativity. Special relativity says the fastest anything can move, including information about where a particle is located, is 299,792,458 m/s. This speed is known as the speed of light or the speed of causality. If the two charged particles were 299,792,458 meters apart and you nudged one of them, the other particle shouldn't know about that for an entire second.
- **Spectroscopy:** Simply put, the Schrödinger equation gets pretty close to calculating the correct energy for atomic states, but it is still wrong.

11.2   The Uncertainty Principle Part 4

In Chap. 6, we introduced the uncertainty principle. To summarize, the uncertainty principle says that if two observables are incompatible, we cannot simultaneously know the values of both observables. More specifically, measuring one observable puts the system in a superposition of the basis set for the other incompatible observable, so we will have a probabilistic result for the value of the second observable. The most famous uncertainty principle, which we talked about in Sect. 6.5, is the Heisenberg Uncertainty Principle named after the German physicist Werner Heisenberg. It is

$$(\Delta x)(\Delta p) \geq \frac{\hbar}{2} \tag{11.1}$$

where Δx is the uncertainty (or range of possible measurements) in the position of the particle and Δp is the uncertainty (or range of possible measurements) in the momentum of the particle. The greater than or equal sign is important as well. For some systems, the product of the two uncertainties is close to $\hbar/2$ while other systems are not. For example, the hydrogen atom with the electron in the ground state has a range of possible position values that can be calculated to be $\Delta x = \frac{\sqrt{3}}{2}a_0$, where $a_0 = 5.29 \times 10^{-11}$ m is a constant called the Bohr radius, named after Danish physicist Niels Bohr. The range of measurements for momentum can be calculated as well, $\Delta p = \frac{\hbar}{a_0}$. The product of the two is $\frac{\sqrt{3}}{2}\hbar$. Notice this is larger than the minimum value of $\frac{\hbar}{2}$.

In quantum mechanics, every pair of incompatible observables has an uncertainty principle. An important uncertainty principle for this discussion is the energy-time uncertainty principle, which is mathematically written as

$$(\Delta E)(\Delta t) \geq \frac{\hbar}{2}. \tag{11.2}$$

This is a very useful uncertainty principle! It says that a state that only exists for a short time cannot have a definite energy. This is why an excited state in an atom has a natural linewidth (Chap. 3). The lifetime of an excited state tells us how long, on average, an electron stays in that state, see Fig. 3.6 and Eq. 3.7. Some electrons move to a lower energy state quickly, while some hang around for a while. This uncertainty principle helps us understand why the lifetime is connected to the natural linewidth of a state, see Eq. 3.8 and Fig. 3.8. In fact, the summary at the end of Chap. 3.3 is stating in words this uncertainty principle!

Let's work with Eq. 3.8. For ease, here is the equation again:

$$\tau = \frac{1}{\Gamma}. \tag{11.3}$$

τ is the lifetime of an excited state, but this is a characteristic time, see the graph in Fig. 11.1. Some states decay quickly ($t < \tau$) while others decay more slowly ($t > \tau$). There is a range or uncertainty in the time it takes for that state to decay. So, let's call the uncertainty in time τ: $\Delta t = \tau$. Now let's think about the energy of the excited state, see the energy level diagram in Fig. 11.1. The energy of the state is centered at the resonance frequency, but there is a width to that resonance. We call that width the natural linewidth. In energy units, the width is $\hbar\Gamma$. This is the range or uncertainty of the excited state: $\Delta E = \hbar\Gamma$. Let's multiply the two uncertainties

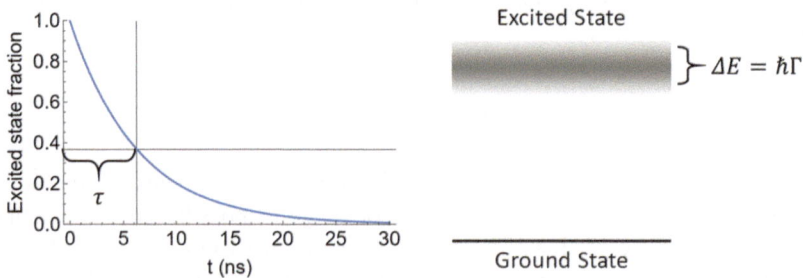

Fig. 11.1 An electron in an excited state decays with a characteristic lifetime τ. That lifetime is related to the linewidth of the excited state by Eq. 11.3. If τ is small, Γ is large. Conversely, if τ is large, Γ is small

together and compare it to the energy-time uncertainty principle, Eq. 11.2:

$$(\Delta E)(\Delta t) = \hbar \, \Gamma \tau \text{ (for an electron in an excited state)} \qquad (11.4)$$

But, Eq. 11.3 tells us that $\Gamma \tau = 1$, so we find

$$(\Delta E)(\Delta t) = \hbar \text{ (for an electron in an excited state)} \qquad (11.5)$$

This is really neat! We now know why a spectral feature has a width and an excited state has a lifetime: time and energy are incompatible observables. While that was fun, you might be thinking, "Why is this in Chap. 11, a chapter devoted to the issues and problems in quantum mechanics?" We will need it in Sect. 11.4.2 to understand something called vacuum fluctuations. But first, we need to discuss one more concept: **antimatter**.

11.3   Antimatter

The Famous Equation

In special relativity, Einstein showed us that energy and mass are the same thing:

$$E = mc^2$$

This equation is for a particle at rest. If we could somehow convert a proton, which has a mass of $1.6726219 \times 10^{-27}$ kg, to energy, we would have:

$$E = mc^2 = 938.27 \times 10^6 \, \text{eV} = 938.27 \, \text{MeV}$$

That is a lot of energy![a] This equation also provides us with a brand new unit for mass. The mass of the proton is $938.27 \, \text{MeV}/c^2$.

---

[a]Converting mass to energy is the basic idea behind nuclear reactors.

In nature, there are a number of conservation laws that seem to always be true. For example, conservation of energy tells us that for an isolated system (i.e., nothing comes in or out), the total energy of the system is a constant value. Other conserved quantities include charge and angular momentum. As we will discuss in the next section, atom number is not a conserved quantity. For example, we can, and have, created particles from high-energy photons. This process is called **pair production**. To accomplish this, we send a high-energy photon towards a nucleus. The nucleus "nudges" the photon a little bit and, if the photon has enough energy, it transforms into an electron and a positron.[4] Before the transformation, the photon existed while the electron and positron did not. After the transformation, the photon no longer exists.

A positron, which we introduced at various times throughout the book, is known as the antimatter partner to the electron. It has the same mass and spin as the electron, but it has opposite (positive) charge. Conservation laws are really helpful at telling us the properties of the positron. Since the photon has spin 1 and the electron has a spin of 1/2, conservation of spin tells us the positron also has a spin of 1/2. A photon has no charge. Likewise, the charges of the electron and positron are the same magnitude, but with opposite signs. This is called conservation of charge. Since the photon has no electric charge, if the photon is to be converted into particles, those particles must together have no charge. There are other differences between the electron and positron as well, but we aren't going to go through all of them.[5] If a positron collides with an electron, the two particles "annihilate" each other back into photons. Before the annihilation, the electron and positron existed while the photons did not. After the annihilation, the electron and positron no longer exist.

Other examples of pair production include creating a muon and an antimuon (see Sect. 11.5), a proton and an antiproton, and a neutron and an antineutron. For this book, the exact details of antimatter don't really matter. What matters is that these antimatter particles can exist, and we have created a lot of antimatter in accelerators or through studying β^+ decay. If an antimatter particle collides with its matter partner, they will annihilate each other creating photons.[6]

Fun Fact

The nuclei of an atom can also have excited states similar to the electrons in the atom. These nuclear excited states are much higher energy compared to electron excited states. Almost all of these excited states emit a high energy photon called an x-ray or a gamma ray to transition back to the nuclear ground state. However, there are nuclei that emit matter and antimatter pairs to transition back to the nuclear ground state! As an example, oxygen-16 has an excited state about 6050×10^3 eV above the nuclear ground state.[b] When

(continued)

---

[4] The nucleus is the external interaction needed to start the process. As a general rule, you always need something acting on the system to initiate such a process.

[5] If you'd like to explore more differences, search the internet for parity. Parity is a concept that is easy to say but hard to understand. Electrons and positrons have opposite parity. Fun fact: we used to think parity was a conserved quantity. However, in 1956, Chinese American physicist Tsung-Dao Lee and Chinese physicist Chen-Ning Yang proposed a theory that the weak force does not conserve parity. This idea was experimentally confirmed by Chinese American physicist Chien-Shiung Wu and her collaborators.

[6] For completeness, other particles besides photons can also be created during annihilation.

an oxygen-16 nucleus is in this excited state, it would violate conservation of angular momentum if it did decay to the nuclear ground state by emitting a photon. So, instead the nucleon gets rid of the energy by emitting an electron and positron in order for the nucleus to transition back to the nuclear ground state.

---

[b]For comparison, a 450 nm wavelength photon has an energy of 2.75 eV. So, this nuclear excited state has over 2 million times more energy!

11.4   Going from Quantum Mechanics to Quantum Field Theory

11.4.1 Remove the Conservation of Particle Number Constraint

Reminder
Quantum mechanics conserves particle number. However, unstable atoms decay, and we create and destroy particles all the time at particle accelerators.

To fix this incompatibility with nature (i.e., the theory says no, but experiment says yes), physicists had to remove the constraint of conservation of particle number from their models. Particle number is clearly not conserved in nature, so we should try to make a model that doesn't conserve particle number. However, this causes a really interesting problem. The Schrödinger equation models an individual particle. What do we do if the particle doesn't always exist? The solution was to form a mathematical model that doesn't try to model a particle, but something called a field. But what is a field? The easiest way to understand a field is with a few examples. Figure 11.2 is a temperature field of the minimum temperatures across the Nordic countries Norway, Sweden, and Finland from January 27, 1999. The region had an unusually cold day.[7] For reference, $-50°$ C $= -58°$ F and $0°$ C $= +32°$ F. This type of field is known as a scalar field. Every point on the map has a temperature. If I told you a coordinate, you could tell me the temperature at that coordinate. In other words, this scalar field is really some function $T(x, y)$. If you knew the function, you could quickly find the temperature for any coordinate values of x and y.

Figure 11.3 is another type of field known as a vector field. This is a plot that shows the speed of wind in a simulated tornado. This is, again, a function, but the answer that you get from this function is a vector. It tells you both the speed of the

---

[7] Det här är för kallt!

Fig. 11.2 A scalar field showing the minimum temperatures across the Nordic countries Norway, Sweden, and Finland on January 27, 1999. The data used to make this plot was taken from the Copernicus Climate Change Service, Climate Data Store, see Reference [1]

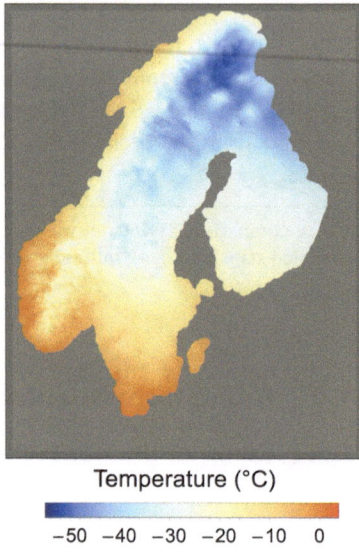

Temperature (°C)

−50 −40 −30 −20 −10   0

wind and the direction. Other vector fields you might have heard about are electric fields and magnetic fields.

Instead of following around a particle, quantum field theory keeps track of a field. Particles are "excitations" of the fields. As an analogy, let's think about the standing waves on a quantum mechanical string, see Fig. 6.2. The background field in this

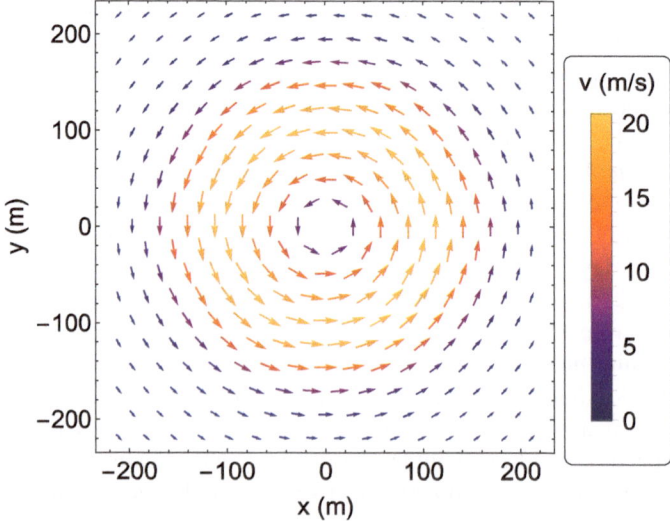

Fig. 11.3 A (simulated) vector field showing the wind speed of a tornado that has a diameter of about 300 m and a top speed of about 20 m/s. This type of plot gives us both direction (the flow is counterclockwise) and the speed of the air

analogy would be the state with a single loop. The particle would be an excitation of this field, or the state with two loops. A quantum particle can go into the field (in our analogy, the system went from two loops to one) or come out of it (one loop to two). Consider a free neutron moving through space. We know from Chap. 10 that a free neutron is unstable and will decay into a proton, an electron, and an anti-electron neutrino with a half-life of about 11 minutes. From a mathematical viewpoint, that neutron is described as moving through a quantum field. When the neutron decays, the neutron goes into the field and a proton, electron, and anti-electron neutrino come out of the field.[8]

> **Important Summary**
> Removing the constraint on conservation of particle number forced physicists to develop a model that doesn't rely on a particle always existing. Instead, quantum field theory models fields where particles can go into or out of these fields.

11.4.2   Vacuum Fluctuations

Vacuum fluctuations are temporary changes in the energy of a point in space due to the energy-time uncertainty principle, Eq. 11.2. Removing the constraint on conservation of particle number combined with the energy-time uncertainty principle, results in bizarre behavior. Quantum field theory predicts that particles are continually popping out of and returning into the field. Interestingly, this behavior is not forced; it naturally comes from removing conservation of particle number and including the energy-time uncertainty principle in the model! As a rough analogy, quantum field theory says these particles are excitations of fields similar to exciting a standing wave from one mode to another. This excited state will not last forever, but decay back into the field after a characteristic time Δt.

Think about that for a moment. According to quantum field theory, there is no such thing as empty space. Space is constantly being filled with particles that pop out of and decay back into fields. This might disturb you a bit. Einstein's famous equation says that $E = mc^2$. Reading this equation, we would say that mass and energy are the same thing just like laser frequency, wavelength, and photon energy are all the same thing.

So, doesn't this violate conservation of energy? Particles that have mass are popping out of the field. Mass is energy, so energy that didn't exist now exists. However, the energy-time uncertainty principle actually allows nature to seemingly violate conservation of energy, but only for a very short amount of time. The more

---

[8] To be completely correct, only a quark inside the neutron goes into the field and another quark comes out, but we haven't talked about quarks yet.

massive the particle that pops into existence (a big ΔE), the quicker that particle must return to the field (a small Δt). Removing conservation of particle number combined with the energy-time uncertainty principle allows this odd behavior to happen. It is important to reiterate that this behavior is a natural consequence and not something forced into the math. The temporarily existing particles are known as **virtual particles**.

These fluctuations can be used to explain why an electron in an excited state decays back to the ground state. These virtual particles, when they exist, push on the atom from the outside. This is the external interaction that causes the electron to fall back to the ground state. Remarkably, theorists can use these vacuum fluctuations to calculate the lifetime of an excited state, and their math correctly predicts the lifetime of excited states.

These virtual particles also explain why the mass of a proton or neutron is so much heavier than the three quarks that make them up. Inside the proton (or neutron), the three quarks are bound tightly together by the strong nuclear force, mediated by virtual particles known as gluons. We will discuss quarks and gluons more in Sect. 11.5. Because of this strong binding, quantum field theory predicts that virtual particles are continuously generated and annihilated within the field. The time-averaged mass of the proton thus includes not only the mass of the three quarks but also the contributions from these virtual particles. Ready to have your mind blown: 99% of the mass of a proton (or neutron) comes from particles popping into and out of the field. That means that 99% of your mass has now gone back into the field and been replaced with new particles from the field.[9]

I recognize that this all probably seems very, very odd. But, every experiment seems to confirm that these surprising predictions are true.

11.4.3 Speed of Causality and Feynman Diagrams

Einstein's theory of special relativity put a speed limit on the universe. Nothing, including information, can travel faster than the speed of light. However, the Schrödinger equation has instantaneous information transfer. How do we solve this problem of instantaneous information transfer?

The answer is **force carriers**. These force carriers are particles that temporarily come from the field to transmit information from one particle to another. These ideas are nicely illustrated in Feynman diagrams, which are named after the American physicist Richard Feynman.

---

[9] For completeness, there are other ways to think about this extra mass. Some physicists prefer to use something called binding energy or simply "fluctuations of the field" to account for the extra mass. All of these ideas are mathematically correct. I have found that virtual particles are a very accessible way for first-time learners to think about the world of the super small. They will also be helpful when we get to force carriers, which are virtual particles, in the next section.

How to read a Feynman diagram:

- Time goes from the bottom to the top.[c] The horizontal axis looks like it should describe a spatial coordinate, but it doesn't. The horizontal axis doesn't really mean anything.
- Particles are represented using straight lines with arrows on them. Matter has arrows that point in the same direction as time (up) while antimatter has arrows that point in the opposite direction as time (down).
- Force carriers are wavy lines.
- Particle creation and annihilation occur at the vertices.

---

[c]Some physicists like to rotate their Feynman diagrams so that time goes left to right instead of bottom to top.

A Feynman diagram is not just a wonderful way to visualize a quantum mechanical process. Each diagram is actually a visual representation of a complicated mathematical equation from quantum field theory. An example Feynman diagram can be seen in Fig. 11.4. In this Feynman diagram, we have two electrons that exchange a "virtual photon," which can exist for a short amount of time because of the energy-time uncertainty principle. That virtual photon comes from the field and does not exist for very long before returning to the field.

The virtual photon transmits information, including the electromagnetic force, from one electron to the other electron. This is how quantum field theory describes the Coulomb interaction! When two electrons are close to each other, the virtual photons that mediate the electromagnetic force between them can exist for a short time due to the energy-time uncertainty principle. Since the virtual photons don't have to exist for a very long time, they can have higher energy, resulting in a strong repulsive interaction between the like charges. As the two electrons move further

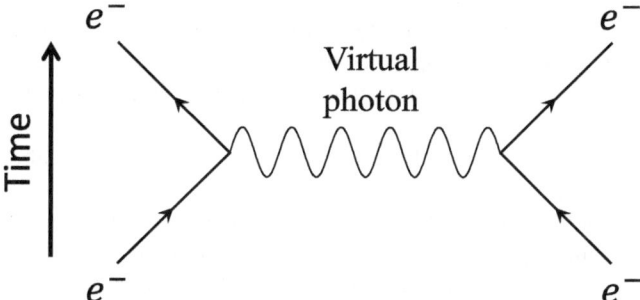

Fig. 11.4 An example of a Feynman diagram. This diagram illustrates a possible interaction between two electrons

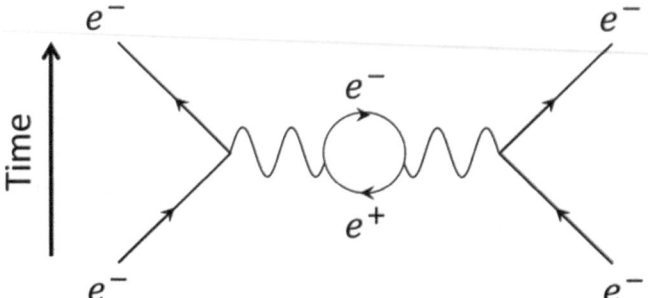

Fig. 11.5 A Feynman diagram depicting how a virtual photon can momentarily split into an electron-positron pair while being exchanged between two electrons

apart, the virtual photons need to exist for a longer time to mediate the interaction, which means they have less energy. Consequently, the force between the two charged particles decreases with distance. This is why, according to quantum field theory, the repulsive or attractive force between two charged particles diminishes as the distance between them increases.

Other things can happen as well. For example, Fig. 11.5 describes how a virtual photon can split into an electron-positron pair when traveling between the two charged particles. This doesn't happen very often compared to the simple exchange of a virtual photon (the electron-positron pair have much more energy, so the energy-time uncertainty principle makes it harder for this to happen), but when the electron-position pair pop into existence from the field, they have a real impact on the system.

Consider an electron in an atom. That electron is constantly interacting with the protons in the nucleus via virtual photons. Every once in a while, that virtual photon breaks apart into an electron-positron pair similar to Fig. 11.5 (just replace one of the electrons with a proton). The effect of the electron-positron pair is that the transition frequency between any two states is shifted by just a little bit, but this small shift is still big enough for us to experimentally measure!

There are other things besides virtual photons and electron-positron pairs that can happen, and theorists calculate the effect that all of these different scenarios will have on an atom. Adding up all of the different possible scenarios produces a single overall shift to the transition frequency. Experimentalists then go measure the transition frequency.

Quantum field theory uses all these ideas to make predictions about the atom. Remarkably, these predictions have almost always been confirmed by experiment. In fact, the theory has been confirmed so many times that if there is a disagreement between theory and experiment, we all assume someone made an error.

There are two major subfields to quantum field theory: quantum electrodynamics and quantum chromodynamics. Quantum electrodynamics (QED) is a mathematical framework that describes interactions between charged particles and light (including virtual photons). Quantum chromodynamics (QCD) is a mathematical framework that describes the interactions inside the nucleus including the interactions of the quarks inside the proton and neutron.

11.4.4 One More Thing

You may have heard some form of the expression, "Particle accelerators make particles." This is a true statement, but we now have the background to understand how they make new particles. According to quantum field theory, we have virtual particles popping into existence from the field and hanging out for a short amount of time before returning to the field.

Important Reminder
Einstein showed us that energy and mass are the same thing

$$E = mc^2$$

Let's take two protons and give them a ton of kinetic energy. Next, we are going to smash the two protons together. If the kinetic energy of the protons is larger than the energy equivalent of the particle's mass coming from the field ($E = mc^2$), the production of that particle does not violate conservation of energy. In this case, the energy-time uncertainty principle is not the limiting factor, allowing the particle to exist as a real particle. This is exactly what particle accelerators do. Almost all of the particles created in accelerators are unstable and undergo radioactive decay, but they exist long enough for us to determine important properties of the particles, such as mass, charge, and spin.

11.5   The Standard Model of Particle Physics

Quantum field theory is currently the best mathematical model that we have to describe the world of the super small. The conceptual model is called the Standard Model of Particle Physics, or just the Standard Model for short, see Fig. 11.6.

A Little History Before the 1960s, the world of particle physics was really confusing. Physicists at particle accelerators were creating hundreds of new particles. It was a little overwhelming; could there really be hundreds upon hundreds of elementary particles? There were so many new particles that in 1956, an American physicist named Robert Oppenheimer coined the term "subnuclear zoo."

In 1964, American physicists Murray Gell-Mann and George Zweig independently realized that the whole crazy picture could be simplified to just a few elementary particles: quarks. Despite particle accelerators seemingly finding new

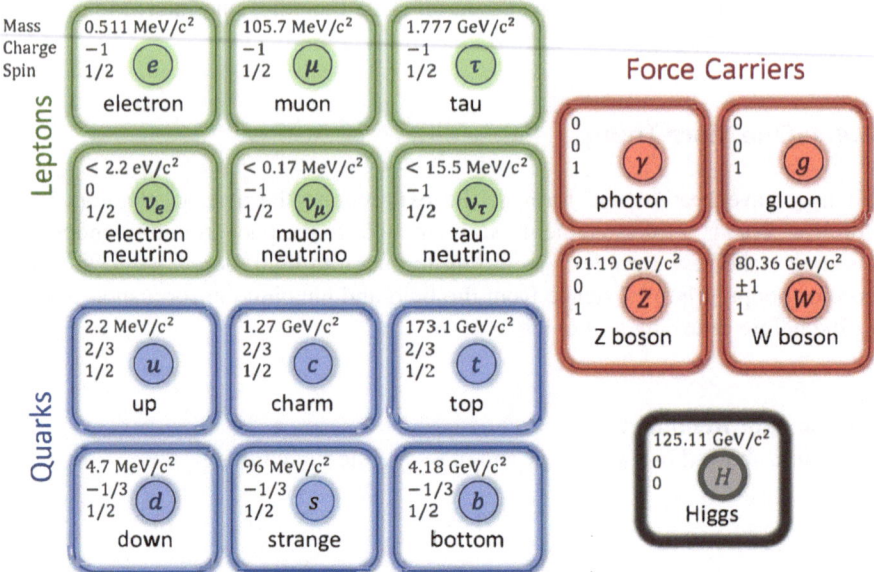

Fig. 11.6 The standard model of particle physics

particles almost every week, what they figured out was that all of the particles were made from twelve elementary particles and their antimatter counterparts. Six of these are now called the up quark, the down quark, the charm quark, the strange quark, the top quark, and the bottom quark,[10] and all of their antimatter counterparts.

At the moment, our best conceptual model of the world of the super small says that everything in the universe is made up of the elementary particles shown in Fig. 11.6. There are four major groupings in the Standard Model.

- **Leptons** are elementary particles that do not experience the strong force and include electrons, muons, taus, neutrinos, and all their antimatter counterparts. They are colored green in Fig. 11.6.
- **Quarks** are elementary particles that experience all fundamental forces, including the strong force. Each quark has an antimatter counterpart. Any particle made from quarks, including protons and neutrons, is called a hadron. They are colored blue in Fig. 11.6.
- **Force Carriers** are particles that mediate the fundamental forces. They include the photon (electromagnetic force), W and Z bosons (weak force), gluons (strong force), and the hypothetical graviton (gravity). They are colored red in Fig. 11.6.
- **Higgs Boson** is an elementary particle associated with the Higgs field, which gives mass to other particles. It is colored black in Fig. 11.6.

---

[10] Some of these names are odd, but it was the 1960s and 1970s . . . .

Protons are composed of 2 up quarks and 1 down quark. A neutron is made up of 1 up quark and 2 down quarks. Helium-4, which has 2 protons, 2 neutrons, and 2 electrons, is actually made up of 6 up quarks, 6 down quarks, and 2 electrons. All of the particles from the subnuclear zoo were made from the six quarks and their antimatter partners. For example, one of the many particles from the zoo is known as the charmed sigma particle, which is actually composed of 2 up quarks and 1 charm quark. Another particle from the zoo is the pion, which is composed of 1 up quark and 1 anti-down quark (the antimatter version of the down quark).

The four elementary particles colored red are the force carriers. Even though the gluon is listed once, there are actually 8 different types of gluons. All 8 of them are carriers for the strong force, which transmit information between quarks. The photon is the force carrier that transmits information between charged particles. Finally, the Z boson and the two W bosons are force carriers for the weak nuclear force, which is the force responsible for β^- decay, β^+ decay, and electron capture. Also, notice how massive the W and Z bosons are. This is why the weak force is such a short range force. The uncertainty principle tells us that those force carriers simply can't exist for very long making their range very, very short.

We can now be more precise in what happens during β^- decay, see Fig. 11.7. During this process, a down quark in the neutron is transformed into an up quark, resulting in the neutron becoming a proton. This transformation is facilitated by the emission of a W^- boson, which quickly decays into an electron and an anti-electron neutrino. The proton remains in the nucleus, while the electron and anti-electron neutrino are emitted from the atom.

Once again, I admit that all of this seems a little ... out there. But everything we have talked about so far really just naturally falls out of the math when we remove or add the restrictions that we talked about in Sect. 11.1. More importantly, the math behind this conceptual model seems to do an amazing job both modeling and predicting the world of the super small.

The final fundamental particle in the Standard Model is the Higgs boson, named after British physicist Peter Higgs. Simply put, the Higgs field interacts with all particles that have mass. Particles with mass are constantly interacting with the

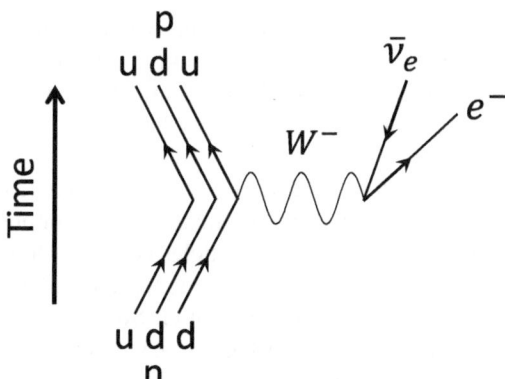

Fig. 11.7 A Feynman diagram showing how a neutron decays into a proton. A neutron is composed of 2 down quarks and 1 up quark. One of the down quarks is transformed into an up quark. This transformation is facilitated by the emission of a W^- boson, which quickly decays into an electron and an anti-electron neutrino

Higgs field through the Higgs boson. This interaction is, according to the Standard Model, the thing that gives the particles mass. This was the last particle to be discovered by particle physicists. The Higgs boson itself was theoretically predicted back in 1964 and finally created in a particle accelerator in 2012. Peter Higgs and Belgian physicist François Englert won the Nobel Prize for the Higgs boson in 2013.

11.6   So, What's Next?

This is exactly the question physicists are always asking! We want to understand nature. To do so, we develop models and test the models to make sure they accurately describe nature. There are different ways to test the Standard Model, and Part 1 of this book discusses how atomic and nuclear physicists use spectroscopy as a tool to test these models. The Standard Model of Particle Physics, which is mathematically described by quantum field theory, is our current best model for the world of the super small. The theory has been tested over and over again, and it has succeeded almost every single time. However, we know the theory is not complete. There are plenty of things in the universe that are not included in the Standard Model. That means that we might be able to create a more complete theory. Here is a list of some of the phenomena that are not in the Standard Model.

Gravity Believe it or not, quantum field theory does not include gravity. Einstein's theory of general relativity is our best model of gravity. General relativity has predicted fascinating phenomena such as gravitational lensing (the theoretically predicted and experimentally measured phenomenon of light bending around massive astronomical objects like large stars and galaxies) and gravitational waves (the theoretically predicted and experimentally measured phenomenon that creates "ripples" in space when two black holes merge; quite literally space is compressed just a little bit and this compression wave ripples out to be detected on earth). Like quantum field theory, general relativity is an incredibly successful theory. Interestingly, the two theories are incompatible with one another. There are a number of reasons why, but one of the big reasons is that general relativity requires "spacetime" to be smooth and continuous while quantum field theory discretizes the fields. In other words, quantum says no to smooth and continuous while general relativity says no to discrete and bumpy. Quantum field theory also needs a force carrier to transmit gravitational information between two objects with mass, which we call a graviton. However, we have yet to create and measure a graviton in the lab.

Where Is All the Antimatter? Quantum field theory predicts there should be about equal parts matter and antimatter. For example, Fig. 11.8 shows a photon that has enough energy to create particles (we have done this at accelerators!). The particles that are created are always a matter/antimatter pair. However, when we look out in the universe, we only see matter. So, where is all the antimatter? We don't know. This problem is called baryon asymmetry. This isn't a bad thing. If the universe was equal parts matter and antimatter, they would have annihilated each other. So,

Fig. 11.8 Everything we
have done in the lab produces
equal amounts of matter and
antimatter, like this photon
making an electron and a
positron

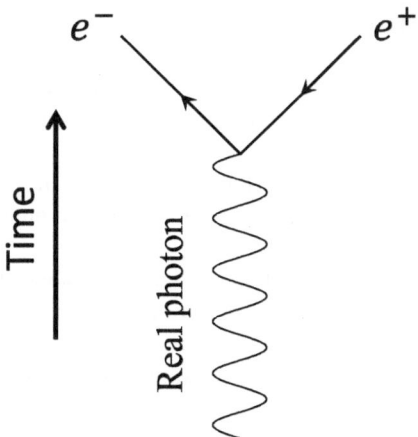

in a way, this is one experimental measurement that the Standard Model failed to
predict. The Standard Model predicts equal amounts of matter and antimatter, while
observations show a universe dominated by matter.

Why Do Neutrinos Have Mass? This one is short. The Standard Model says that
neutrinos don't have mass. However, they do. This is why I use the phrase "The
theory has been tested over and over again, and it has succeeded almost every single
time." The Standard Model says neutrinos don't have mass, but we have found that
they do.

What Is Dark Matter and Dark Energy? This one is also short. From astronomical observations, the universe is thought to be about 5% atoms, 26% dark matter, and
69% dark energy. However, we don't even know what dark matter or dark energy
is . . . so it is a little hard to include them in a model. Still, it is kinda weird that our
most successful model ever tested (quantum field theory) doesn't know how to deal
with 95% of the universe.

Even though quantum field theory has been incredibly successful, there is
still a lot to learn! For physicists, this is the best thing ever. There have been
some incredibly interesting and ingenious ideas to make the Standard Model more
complete. Some of these ideas include string theory, supersymmetry, and loop
quantum gravity. However, no one has come up with any way to definitively test
if any of these ideas are correct. But this is what makes everything so exciting! We
still have more to learn, more to understand, and more to explore. To learn more and
push our understanding always onward, we use the most important equation in all
of physics:

$$\text{questions} + \text{repetition} + \text{critical thinking} = \text{mastery}$$

Problems

11.1 Explain why the Schrödinger equation is not suitable for describing particles moving at speeds close to the speed of light.

11.2 Describe the process of pair production and explain why it violates the conservation of particle number in quantum mechanics.

11.3 In this chapter, we talked about temperature as an example of a scalar field and the wind velocity of a tornado as an example of a vector field.

(a) What is another example of a scalar field?
(b) What is another example of a vector field?

11.4 In Sect. 11.3, we stated the mass of a proton can be written as $938.27\,\mathrm{MeV}/c^2$.

(a) Show this to be true.
    Hint: $1\,\mathrm{MeV} = 1.602176 \times 10^{-13}\,\mathrm{J}$
(b) The mass of an electron is $9.109384 \times 10^{-31}\,\mathrm{kg}$. Write the mass in units of MeV/c^2.
(c) What is the mass of the charm quark in kilograms?

11.5 Draw a Feynman diagram for β^+ decay. The force carrier will be a W^+ boson.

11.6 Draw a Feynman diagram for electron capture. The force carrier will be a W^+ boson.

11.7 Explain how vacuum fluctuations can cause an electron in an excited state to decay back to its ground state.

11.8 An accelerator accelerates two protons to very high energy. The two protons travel in opposite directions with the same speed and collide. The goal is to create a top quark and an anti-top quark.

(a) What is the minimum energy each proton must have to make a top quark-antiquark pair?
(b) The relationship between the energy of a particle and its velocity comes from special relativity. The formula is

$$E = \frac{mc^2}{\sqrt{1 - \frac{v^2}{c^2}}}$$

Using your answer from part (a), what is the speed of a proton? Express your answer in units of c. For example, an incorrect answer is $v = 0.92c$, or 92% the speed of light.

Fun Fact The Tevatron was a particle accelerator at Fermi National Accelerator Laboratory (Fermilab), located in Batavia, Illinois, near Chicago. It accelerated protons and antiprotons to energies of 980 GeV, producing proton-antiproton collisions with energies of up to 1.96 TeV. Sadly, the Tevatron shut down in 2011 because the Large Hadron Collider (LHC) at CERN, which became operational in 2008, surpassed the Tevatron in terms of energy levels. The LHC can collide protons at energies of 7 TeV per beam (14 TeV in total). CERN is located in Geneva, Switzerland, and it is currently the world's largest particle accelerator. CERN is an acronym for the French name "Conseil Européen pour la Recherche Nucléaire," which translates to "The European Organization for Nuclear Research."

Reference

1. Tveito, O.E., Førland, E.J., Heino, R., Hanssen-Bauer, I., Alexandersson, H., Dahlström, B., Drebs, A., Kern-Hansen, C., Jónsson, T., Vaarby-Laursen E., Westman, Y.: Nordic Temperature Maps DNMI Klima 9/00 KLIMA. Norwegian Meteorological Institute, Oslo (2000). Copernicus Climate Change Service, Climate Data Store, (2021): Nordic gridded temperature and precipitation data from 1971 to present derived from in-situ observations. Copernicus Climate Change Service (C3S) Climate Data Store (CDS). https://doi.org/10.24381/cds.e8f4a10c, Accessed Jan 25, 2024

The Periodic Table of the Elements

Room temperature states of matter

- Gas
- Liquid
- Solid
- Artificially Prepared

1 H Hydrogen																	2 He Helium
3 Li Lithium	4 Be Beryllium											5 B Boron	6 C Carbon	7 N Nitrogen	8 O Oxygen	9 F Fluorine	10 Ne Neon
11 Na Sodium	12 Mg Magnesium											13 Al Aluminum	14 Si Silicon	15 P Phosphorus	16 S Sulfur	17 Cl Chlorine	18 Ar Argon
19 K Potassium	20 Ca Calcium	21 Sc Scandium	22 Ti Titanium	23 V Vanadium	24 Cr Chromium	25 Mn Manganese	26 Fe Iron	27 Co Cobalt	28 Ni Nickel	29 Cu Copper	30 Zn Zinc	31 Ga Gallium	32 Ge Germanium	33 As Arsenic	34 Se Selenium	35 Br Bromine	36 Kr Krypton
37 Rb Rubidium	38 Sr Strontium	39 Y Yttrium	40 Zr Zirconium	41 Nb Niobium	42 Mo Molybdenum	43 Tc Technetium	44 Ru Ruthenium	45 Rh Rhodium	46 Pd Palladium	47 Ag Silver	48 Cd Cadmium	49 In Indium	50 Sn Tin	51 Sb Antimony	52 Te Tellurium	53 I Iodine	54 Xe Xenon
55 Cs Cesium	56 Ba Barium	57 La Lanthanum	72 Hf Hafnium	73 Ta Tantalum	74 W Tungsten	75 Re Rhenium	76 Os Osmium	77 Ir Iridium	78 Pt Platinum	79 Au Gold	80 Hg Mercury	81 Tl Thallium	82 Pb Lead	83 Bi Bismuth	84 Po Polonium	85 At Astatine	86 Rn Radon
87 Fr Francium	88 Ra Radium	89 Ac Actinium	104 Rf Rutherfordium	105 Db Dubnium	106 Sg Seaborgium	107 Bh Bohrium	108 Hs Hassium	109 Mt Meitnerium	110 Ds Darmstadtium	111 Rg Roentgenium	112 Cn Copernicium	113 Nh Nihonium	114 Fl Flerovium	115 Mc Moscovium	116 Lv Livermorium	117 Ts Tennessine	118 Og Oganesson

58 Ce Cerium	59 Pr Praseodymium	60 Nd Neodymium	61 Pm Promethium	62 Sm Samarium	63 Eu Europium	64 Gd Gadolinium	65 Tb Terbium	66 Dy Dysprosium	67 Ho Holmium	68 Er Erbium	69 Tm Thulium	70 Yb Ytterbium	71 Lu Lutetium
90 Th Thorium	91 Pa Protactinium	92 U Uranium	93 Np Neptunium	94 Pu Plutonium	95 Am Americium	96 Cm Curium	97 Bk Berkelium	98 Cf Californium	99 Es Einsteinium	100 Fm Fermium	101 Md Mendelevium	102 No Nobelium	103 Lr Lawrencium

© The Author(s) 2025
W. Raven, *Atomic Physics for Everyone*,
https://doi.org/10.1007/978-3-031-69507-0

245

A Table of the Elements

B

Below is a list of all of the known elements sorted by the number of protons in the nucleus (Z). The last column lists the mass number (A) for all of the stable isotopes. For example, the stable isotope magnesium-25 has 12 protons and 13 neutrons for a total of 25 nucleons.

Z	Name	Symbol	Ground state electron configuration	Mass number for stable isotopes
1	Hydrogen	H	$1s\,^2S_{1/2}$	1, 2
2	Helium	He	$1s^2\,^1S_0$	3, 4
3	Lithium	Li	$[\text{He}]2s\,^2S_{1/2}$	6, 7
4	Beryllium	Be	$[\text{He}]2s^2\,^1S_0$	9
5	Boron	B	$[\text{He}]2s^22p^1\,^2P_{1/2}$	10, 11
6	Carbon	C	$[\text{He}]2s^22p^2\,^3P_0$	12, 13
7	Nitrogen	N	$[\text{He}]2s^22p^3\,^4S_{3/2}$	14, 15
8	Oxygen	O	$[\text{He}]2s^22p^4\,^3P_2$	16, 17, 18
9	Fluorine	F	$[\text{He}]2s^22p^5\,^2P_{3/2}$	19
10	Neon	Ne	$[\text{He}]2s^22p^6\,^1S_0$	20, 21, 22
11	Sodium	Na	$[\text{Ne}]3s^1\,^2S_{1/2}$	23
12	Magnesium	Mg	$[\text{Ne}]3s^2\,^1S_0$	24, 25, 26
13	Aluminium	Al	$[\text{Ne}]3s^23p^1\,^2P_{1/2}$	27
14	Silicon	Si	$[\text{Ne}]3s^23p^2\,^3P_0$	28, 29, 30
15	Phosphorus	P	$[\text{Ne}]3s^23p^3\,^4S_{3/2}$	31
16	Sulfur	S	$[\text{Ne}]3s^23p^4\,^3P_2$	32, 33, 34, 36
17	Chlorine	Cl	$[\text{Ne}]3s^23p^5\,^2P_{3/2}$	35, 37
18	Argon	Ar	$[\text{Ne}]3s^23p^6\,^1S_0$	36, 38, 40
19	Potassium	K	$[\text{Ar}]4s^1\,^2S_{1/2}$	39, 40, 41
20	Calcium	Ca	$[\text{Ar}]4s^2\,^1S_0$	40, 42, 43, 44, 46, 48

(continued)

© The Author(s) 2025
W. Raven, *Atomic Physics for Everyone*,
https://doi.org/10.1007/978-3-031-69507-0

Z	Name	Symbol	Ground state electron configuration	Stable isotopes
21	Scandium	Sc	$[\text{Ar}]3d^1 4s^2\ ^2D_{3/2}$	45
22	Titanium	Ti	$[\text{Ar}]3d^2 4s^2\ ^3F_2$	46, 47, 48, 49, 50
23	Vanadium	V	$[\text{Ar}]3d^3 4s^2\ ^4F_{3/2}$	50, 51
24	Chromium	Cr	$[\text{Ar}]3d^5 4s^1\ ^7S_3$	50, 52, 53, 54
25	Manganese	Mn	$[\text{Ar}]3d^5 4s^2\ ^6S_{5/2}$	55
26	Iron	Fe	$[\text{Ar}]3d^6 4s^2\ ^5D_4$	54, 56, 57, 58
27	Cobalt	Co	$[\text{Ar}]3d^7 4s^2\ ^4F_{9/2}$	59
28	Nickel	Ni	$[\text{Ar}]3d^8 4s^2\ ^3F_4$	58, 60, 61, 62, 64
29	Copper	Cu	$[\text{Ar}]3d^{10} 4s^1\ ^2S_{1/2}$	63, 65
30	Zinc	Zn	$[\text{Ar}]3d^{10} 4s^2\ ^1S_0$	64, 66, 67, 68, 70
31	Gallium	Ga	$[\text{Ar}]3d^{10} 4s^2 4p^1\ ^2P_{1/2}$	69, 71
32	Germanium	Ge	$[\text{Ar}]3d^{10} 4s^2 4p^2\ ^3P_0$	70, 72, 73, 74, 76
33	Arsenic	As	$[\text{Ar}]3d^{10} 4s^2 4p^3\ ^4S_{3/2}$	75
34	Selenium	Se	$[\text{Ar}]3d^{10} 4s^2 4p^4\ ^3P_2$	74, 76, 77, 78, 80, 82
35	Bromine	Br	$[\text{Ar}]3d^{10} 4s^2 4p^5\ ^2P_{3/2}$	79, 81
36	Krypton	Kr	$[\text{Ar}]3d^{10} 4s^2 4p^6\ ^1S_0$	78, 80, 82, 83, 84, 86
37	Rubidium	Rb	$[\text{Kr}]5s^1\ ^2S_{1/2}$	85, 87
38	Strontium	Sr	$[\text{Kr}]5s^2\ ^1S_0$	84, 86, 87, 88
39	Yttrium	Y	$[\text{Kr}]4d^1 5s^2\ ^2D_{3/2}$	89
40	Zirconium	Zr	$[\text{Kr}]4d^2 5s^2\ ^3F_2$	90, 91, 92, 94, 96
41	Niobium	Nb	$[\text{Kr}]4d^4 5s^1\ ^6D_{1/2}$	93
42	Molybdenum	Mo	$[\text{Kr}]4d^5 5s^1\ ^7S_3$	92, 94, 95, 96, 97, 98
43	Technetium	Tc	$[\text{Kr}]4d^5 5s^2\ ^6S_{5/2}$	N/A
44	Ruthenium	Ru	$[\text{Kr}]4d^7 5s^1\ ^5F_5$	96, 98, 99, 100, 101, 102, 104
45	Rhodium	Rh	$[\text{Kr}]4d^8 5s^1\ ^4F_{9/2}$	103
46	Palladium	Pd	$[\text{Kr}]4d^{10}\ ^1S_0$	102, 104, 105, 106, 108, 110
47	Silver	Ag	$[\text{Kr}]4d^{10} 5s^1\ ^2S_{1/2}$	107, 109
48	Cadmium	Cd	$[\text{Kr}]4d^{10} 5s^2\ ^1S_0$	106, 108, 110, 111, 112, 113, 114, 116
49	Indium	In	$[\text{Kr}]4d^{10} 5s^2 5p^1\ ^2P_{1/2}$	113, 115
50	Tin	Sn	$[\text{Kr}]4d^{10} 5s^2 5p^2\ ^3P_0$	112, 114, 115, 116, 117, 118, 119, 120, 122, 124
51	Antimony	Sb	$[\text{Kr}]4d^{10} 5s^2 5p^3\ ^4S_{3/2}$	121, 123
52	Tellurium	Te	$[\text{Kr}]4d^{10} 5s^2 5p^4\ ^3P_2$	120, 122, 123, 124, 125, 126, 128, 130
53	Iodine	I	$[\text{Kr}]4d^{10} 5s^2 5p^5\ ^2P_{3/2}$	127
54	Xenon	Xe	$[\text{Kr}]4d^{10} 5s^2 5p^6\ ^1S_0$	124, 126, 128, 129, 130, 131, 132, 134, 136
55	Cesium	Cs	$[\text{Xe}]6s^1\ ^2S_{1/2}$	133

(continued)

Z	Name	Symbol	Ground state electron configuration	Stable isotopes
56	Barium	Ba	$[Xe]6s^2\ {}^1S_0$	130, 132, 134, 135, 136, 137, 138
57	Lanthanum	La	$[Xe]5d^1 6s^2\ {}^2D_{3/2}$	138, 139
58	Cerium	Ce	$[Xe]4f^1 5d^1 6s^2\ {}^1G_4$	136, 138, 140, 142
59	Praseodymium	Pr	$[Xe]4f^3 6s^2\ {}^4I_{9/2}$	141
60	Neodymium	Nd	$[Xe]4f^4 6s^2\ {}^5I_4$	142, 143, 144, 145, 146, 148, 150
61	Promethium	Pm	$[Xe]4f^5 6s^2\ {}^6H_{5/2}$	N/A
62	Samarium	Sm	$[Xe]4f^6 6s^2\ {}^7F_0$	144, 147, 148, 149, 150, 152, 154
63	Europium	Eu	$[Xe]4f^7 6s^2\ {}^8S_{7/2}$	151, 153
64	Gadolinium	Gd	$[Xe]4f^7 5d^1 6s^2\ {}^9D_2$	152, 154, 155, 156, 157, 158, 160
65	Terbium	Tb	$[Xe]4f^9 6s^2\ {}^6H_{15/2}$	159
66	Dysprosium	Dy	$[Xe]4f^{10} 6s^2\ {}^5I_8$	156, 158, 160, 161, 162, 163, 164
67	Holmium	Ho	$[Xe]4f^{11} 6s^2\ {}^4I_{15/2}$	165
68	Erbium	Er	$[Xe]4f^{12} 6s^2\ {}^3H_6$	162, 164, 166, 167, 168, 170
69	Thulium	Tm	$[Xe]4f^{13} 6s^2\ {}^2F_{7/2}$	169
70	Ytterbium	Yb	$[Xe]4f^{14} 6s^2\ {}^1S_0$	168, 170, 171, 172, 173, 174, 176
71	Lutetium	Lu	$[Xe]4f^{14} 5d^1 6s^2\ {}^2D_{3/2}$	175, 176
72	Hafnium	Hf	$[Xe]4f^{14} 5d^2 6s^2\ {}^3F_2$	174, 176, 177, 178, 179, 180
73	Tantalum	Ta	$[Xe]4f^{14} 5d^3 6s^2\ {}^4F_{3/2}$	180, 181
74	Tungsten	W	$[Xe]4f^{14} 5d^4 6s^2\ {}^5D_0$	182, 183, 184, 186
75	Rhenium	Re	$[Xe]4f^{14} 5d^5 6s^2\ {}^6S_{5/2}$	185, 187
76	Osmium	Os	$[Xe]4f^{14} 5d^6 6s^2\ {}^5D_4$	184, 186, 187, 188, 189, 190, 192
77	Iridium	Ir	$[Xe]4f^{14} 5d^7 6s^2\ {}^4F_{9/2}$	191, 193
78	Platinum	Pt	$[Xe]4f^{14} 5d^9 6s^1\ {}^3D_3$	190, 192, 194, 195, 196, 198
79	Gold	Au	$[Xe]4f^{14} 5d^{10} 6s^1\ {}^2S_{1/2}$	197
80	Mercury	Hg	$[Xe]4f^{14} 5d^{10} 6s^2\ {}^1S_0$	196, 198, 199, 200, 201, 202, 204
81	Thallium	Tl	$[Xe]4f^{14} 5d^{10} 6s^2 6p^1\ {}^2P_{1/2}$	203, 205
82	Lead	Pb	$[Xe]4f^{14} 5d^{10} 6s^2 6p^2\ {}^3P_0$	204, 206, 207, 208
83	Bismuth	Bi	$[Xe]4f^{14} 5d^{10} 6s^2 6p^3\ {}^4S_{3/2}$	209
84	Polonium	Po	$[Xe]4f^{14} 5d^{10} 6s^2 6p^4\ {}^3P_2$	N/A
85	Astatine	At	$[Xe]4f^{14} 5d^{10} 6s^2 6p^5\ {}^2P_{3/2}$	N/A

(continued)

Z	Name	Symbol	Ground state electron configuration	Stable isotopes
86	Radon	Rn	$[Xe]4f^{14}5d^{10}6s^26p^6\ ^1S_0$	N/A
87	Francium	Fr	$[Rn]7s^1\ ^2S_{1/2}$	N/A
88	Radium	Ra	$[Rn]7s^2\ ^1S_0$	N/A
89	Actinium	Ac	$[Rn]6d^17s^2\ ^2D_{3/2}$	N/A
90	Thorium	Th	$[Rn]6d^27s^2\ ^3F_2$	232
91	Protactinium	Pa	$[Rn]5f^26d^17s^2\ ^4K_{11/2}$	231
92	Uranium	U	$[Rn]5f^36d^17s^2\ ^5L_6$	234, 235, 238
93	Neptunium	Np	$[Rn]5f^46d^17s^2\ ^6L_{11/2}$	N/A
94	Plutonium	Pu	$[Rn]5f^67s^2\ ^7F_0$	N/A
95	Americium	Am	$[Rn]5f^77s^2\ ^8S_{7/2}$	N/A
96	Curium	Cm	$[Rn]5f^76d^17s^2\ ^9D_2$	N/A
97	Berkelium	Bk	$[Rn]5f^97s^2\ ^6H_{15/2}$	N/A
98	Californium	Cf	$[Rn]5f^{10}7s^2\ ^5I_8$	N/A
99	Einsteinium	Es	$[Rn]5f^{11}7s^2\ ^4I_{15/2}$	N/A
100	Fermium	Fm	$[Rn]5f^{12}7s^2\ ^3H_6$	N/A
101	Mendelevium	Md	$[Rn]5f^{13}7s^2\ ^2F_{7/2}$	N/A
102	Nobelium	No	$[Rn]5f^{14}7s^2\ ^1S_0$	N/A
103	Lawrencium	Lr	$[Rn]5f^{14}7s^27p^1\ ^2P_{1/2}$	N/A
104	Rutherfordium	Rf	$[Rn]5f^{14}6d^27s^2\ ^3F_2$	N/A
105	Dubnium	Db	$[Rn]5f^{14}6d^37s^2\ ^4F_{3/2}$	N/A
106	Seaborgium	Sg	$[Rn]5f^{14}6d^47s^2\ ^5D_0$	N/A
107	Bohrium	Bh	$[Rn]5f^{14}6d^57s^2\ ^6S_{5/2}$	N/A
108	Hassium	Hs	$[Rn]5f^{14}6d^67s^2\ ^5D_4$	N/A
109	Meitnerium	Mt	$[Rn]5f^{14}6d^77s^2\ ^4F_{9/2}$	N/A
110	Darmstadtium	Ds	$[Rn]5f^{14}6d^97s^1\ ^3D_3$	N/A
111	Roentgenium	Rg	$[Rn]5f^{14}6d^{10}7s^1\ ^2S_{1/2}$	N/A
112	Copernicium	Cn	$[Rn]5f^{14}6d^{10}7s^2\ ^1S_0$	N/A
113	Nihonium	Nh	$[Rn]5f^{14}6d^{10}7s^27p^1\ ^2P_{1/2}$	N/A
114	Flerovium	Fl	$[Rn]5f^{14}6d^{10}7s^27p^2\ ^3P_0$	N/A
115	Moscovium	Mc	$[Rn]5f^{14}6d^{10}7s^27p^3\ ^4S_{3/2}$	N/A
116	Livermorium	Lv	$[Rn]5f^{14}6d^{10}7s^27p^4\ ^3P_2$	N/A
117	Tennessine	Ts	$[Rn]5f^{14}6d^{10}7s^27p^5\ ^2P_{3/2}$	N/A
118	Oganesson	Og	$[Rn]5f^{14}6d^{10}7s^27p^6\ ^1S_0$	N/A

Transition Rules:

C

The following transition rules (also known as selection rules) must be satisfied for an electron to transition between two atomic states:

General rules for all atoms:

- $\Delta L = \pm 1$: The electronic orbital angular momentum quantum number must change by 1.
- $\Delta S = 0$: The spin quantum number must remain unchanged.
- $\Delta J = -1,\ 0,\ +1;\quad J = 0 \nrightarrow J = 0$: The total electronic angular momentum quantum number must change by 1, 0, or -1. However, transitions between $J = 0 \to J = 0$ are forbidden.
- Parity must change: A transition must occur between states of opposite parity. Parity is indicated by a circle in the term symbol. For example, a state with odd parity has a small circle, like $^1P_1^\circ$. A state with even parity does not have a circle, like $^1S_0$.

For atoms with no nuclear spin:

- $\Delta m_J = -1, 0, +1$: The z-component (projection) of the total electronic angular momentum (m_J) can change by 1, 0, or -1. However, if $\Delta J = 0$, the transition between $m_J = 0 \to m_J = 0$ is forbidden.

For atoms with nuclear spin:

- $\Delta F = -1,\ 0,\ +1;\quad F = 0 \nrightarrow F = 0$: The total atomic angular momentum quantum number including nuclear spin (F) must change by 1, 0, or -1. However, transitions between $F = 0 \to F = 0$ are forbidden.

© The Author(s) 2025
W. Raven, *Atomic Physics for Everyone*,
https://doi.org/10.1007/978-3-031-69507-0

- $\Delta m_F = -1,\ 0,\ +1$: The z-component (projection) of the total atomic angular momentum including nuclear spin (m_F) can change by 1, 0, or -1. However, if $\Delta F = 0$, the transition between $m_F = 0 \rightarrow m_F = 0$ is forbidden.

Glossary

Absorption plot: A plot showing the percentage or fraction of photons absorbed from a laser beam as it passes through a vapor cell, plotted as a function of laser frequency. A dip in a transmission plot corresponds to a peak in the absorption plot.

Angular Momentum: A property of a rotating object that changes when a torque is applied. It is the rotational analog of linear momentum.

Antimatter: Particles that have the same mass as their matter counterparts but opposite charge, as well as opposite parity and spin projection. When an antimatter particle collides with its matter counterpart, they annihilate each other, often creating photons.

Atomic Physics: A field of physics that seeks to understand how the components of an atom interact and behave.

Basis set: The complete set of states for a particular observable in a given system.

Boson: a particle with integer spin. Bosons do not follow the Pauli Exclusion Principle, meaning multiple bosons can occupy the same quantum state.

Bra-ket notation: A common notation used to represent states in quantum mechanics. A "ket" represents a particular state and is written as |state label⟩. A "bra," which is not used in this book but is common in advanced quantum mechanics, is written as ⟨state label|.

Center of gravity: The energy of an atomic state assuming the nucleus has no angular momentum.

Compatible observables: Observables that can be precisely measured simultaneously without affecting each other's measurement. Measuring one observable does not alter the system in a way that impacts the measurement of the other.

Coulomb force: The force between two charged particles, given by $F = k_e \frac{q_1 q_2}{r^2}$, where F is the force, $k_e = 8.987 \times 10^9$ N·m^2/C^2 is Coulomb's constant, q_1 and q_2 are the charges, and r is the distance between them. It describes the attraction or repulsion between charged particles.

Coulomb interaction: The interaction between two charged particles due to their electric charges, governed by the Coulomb force. It is fundamental in understanding the behavior of charged particles in electrostatic situations

© The Author(s) 2025
W. Raven, *Atomic Physics for Everyone*,
https://doi.org/10.1007/978-3-031-69507-0

Degenerate or Degeneracy: When multiple states in a quantum system have the same energy.

Detuning: The difference between the laser frequency and the resonance frequency. When using linear frequency units, the variable representing detuning is the lowercase Greek letter delta: $\delta = f - f_r$, where f is the laser frequency and f_r is the resonance frequency. When using angular frequency units, the variable representing detuning is the uppercase Greek letter delta: $\Delta = \omega - \omega_r$, where $\omega = 2\pi f$ is the frequency of the laser in angular frequency units and $\omega_r = 2\pi f_r$ is the resonance frequency in angular frequency units.

Diffraction: The bending of waves around the corners of an obstacle.

Dispersive element: An optical element that spatially separates light into its spectral components.

Doppler broadening: The widening of a spectral feature caused by the thermal motion of atoms.

Doppler effect: The change in frequency of sound, light, or other waves as the source and observer move toward or away from each other.

Doppler shift: The change in the frequency of light observed by an atom due to the Doppler effect.

Doppler profile: A spectral feature broadened due to the thermal motion of atoms.

Doppler width: The full width at half maximum (FWHM) of a Doppler-broadened spectral feature. The variable Δf_{FWHM} represents the Doppler width.

Electromagnetic force: A fundamental force encompassing both electric and magnetic forces, including the Coulomb force and magnetic forces due to moving charges.

Electromagnetic interaction: The interaction between charged particles due to both electric and magnetic forces, including interactions involving stationary charges (electrostatics) and moving charges.

Electron: A fundamental particle found in an atom outside of the nucleus. An electron has a negative charge of -1.602×10^{-19} C, a mass of 9.109×10^{-31} kg, and no measurable radius. Unlike protons or neutrons, electrons are not composed of smaller particles.

Electron Configuration: The arrangement of electrons in an atom or molecule within designated orbitals around the nucleus. Electron configurations are typically represented using notation that indicates the energy levels, subshells, and the number of electrons in each subshell. For example, the electron configuration of the ground state of carbon, which has six electrons, is $1s^2 2s^2 2p^2$.

Electron Shells: The different energy levels around the nucleus of an atom where electrons are likely to be found. These shells are defined by the principal quantum number n, with each shell corresponding to a specific energy level. Electrons in lower-numbered shells are closer to the nucleus, while those in higher-numbered shells are farther away. Each shell can hold a limited number of electrons, determined by the formula $2n^2$. The arrangement of electrons in these shells follows the rules of quantum mechanics and significantly influences the chemical properties of the atom.

Electron Subshells: Subdivisions of electron shells within an atom, characterized by the quantum number ℓ. Each subshell corresponds to a specific value of ℓ and is labeled using the letters s, p, d, f, etc., corresponding to $\ell = 0, 1, 2, 3$, and so on. Subshells define the shape of the electron's wavefunction and can hold a specific number of electrons: the s subshell can hold 2 electrons, the p subshell can hold 6, the d subshell can hold 10, and the f subshell can hold 14. The arrangement of electrons within these subshells follows the Pauli exclusion principle.

The Equipartition Theorem: A theorem in thermodynamics stating that the energy of a system is equally distributed among all its degrees of freedom.

Fermion: a particle with half-integer spin. Fermions follow the Pauli Exclusion Principle, meaning no two fermions can occupy the same quantum state.

Fine structure splitting: The splitting of atomic energy levels due to the interaction between an electron's orbital angular momentum and spin.

Force:

If the object's mass is constant: An external interaction that causes an object to accelerate (i.e., change velocity). If multiple forces act on an object, the net force causes the object to accelerate. This is Newton's 2nd Law for an object with constant mass: "The sum of all forces acting on an object causes the object to accelerate." This is the most common definition of a force but is a special case of the general definition.

If the object's mass is changing: An external interaction that causes an object's momentum to change. Newton's 2nd Law for momentum states: "The sum of all forces acting on an object causes its momentum to change." This is the general definition of Newton's 2nd law.

The variable F represents force.

Force carriers: Virtual particles that mediate the fundamental forces of nature in quantum field theory. They are bosons and are responsible for transmitting forces between particles. The primary force carriers and the forces they mediate are:

- **Photon**: The force carrier of the electromagnetic force. Photons are massless and mediate interactions between charged particles.
- **W and Z Bosons**: The force carriers of the weak nuclear force. These bosons are massive and are responsible for processes such as beta decay in radioactive materials.
- **Gluons**: The force carriers of the strong nuclear force. Gluons are massless and mediate the force that holds quarks together within protons and neutrons.
- **Graviton** (hypothetical): The proposed force carrier of gravity in quantum gravity theories. Gravitons are hypothesized to be massless and mediate the gravitational force, although they have not been experimentally observed.

Frequency: The number of oscillations per second. The variable f represents frequency. The unit for frequency is 1/second, called Hertz (Hz). Other common units for frequency include megahertz ($1\,\text{MHz} = 1 \times 10^6\,\text{Hz}$), gigahertz ($1\,\text{GHz} = 1 \times 10^9\,\text{Hz}$), and terahertz ($1\,\text{THz} = 1 \times 10^{12}\,\text{Hz}$).

Higgs Boson: An elementary particle associated with the Higgs field, responsible for giving mass to other particles.

Hyperfine level: When a nucleus has angular momentum, an atomic state splits into multiple closely spaced states known as hyperfine levels.

Hyperfine structure splitting: The splitting of energy levels in an atom due to nuclear spin.

Incompatible observables: Observables that cannot be precisely measured simultaneously. Measuring one observable alters the system, making subsequent measurements of the other unpredictable.

Intensity: The intensity of a laser beam is the power of the laser divided by the cross sectional area of the laser beam, denoted as $I = P/A$.

Ion: An atom with a net charge due to an unequal number of protons and electrons. An ion with more protons than electrons is called a positive ion, while an ion with more electrons than protons is called a negative ion.

Ionization Threshold: The minimum energy required to remove an electron from an atom or molecule, thereby ionizing it. Each atom has a unique ionization threshold. Most atoms have multiple ionization thresholds depending on the angular momentum and energy state of the remaining electrons.

Isotope shift: The change in the resonance frequency of a transition between two states in an atom due to a change in the number of neutrons in the nucleus. The variable for isotope shift is $\delta f^{AA'}$, where A is the mass number for one isotope and A' is the mass number of the other isotope.

Kinetic energy: The energy of motion. The variable K represents kinetic energy.

Leptons: Elementary particles that do not experience the strong force, including electrons, muons, taus, neutrinos, and their antimatter counterparts.

Mass number: The total number of protons and neutrons in an atomic nucleus, represented by the variable A.

Molecule: A group of two or more atoms bonded together by chemical bonds.

Momentum: A property of a moving object. For a classical object, momentum is given by $p = mv$, where p is the object's momentum, m is the object's mass, and v is the object's velocity. An object's momentum changes if acted upon by an external force. The unit of momentum is kgm/s.

Natural linewidth: The minimum possible full width at half maximum (FWHM) of a spectral feature. The natural linewidth is a property of a transition, with each transition in an atom having a unique natural linewidth. The variable for natural linewidth is the lowercase Greek letter gamma γ when using linear frequency units and the uppercase Greek letter gamma Γ when using angular frequency units.

Neutral atom: An atom with no net charge, having an equal number of electrons and protons.

Neutron: A particle found inside the nucleus of an atom. A neutron has no charge, a mass of 1.675×10^{-27} kg, and a radius of 0.8×10^{-15} m $= 0.8$ fm. A neutron is composed of three quarks: 1 up quark and 2 down quarks. A free neutron is unstable and undergoes β^- decay with a half life of about 11 minutes.

Neutron number: The total number of neutrons in an atomic nucleus, represented by the variable N.

Nucleon: A term referring to either a neutron or a proton.

Nuclear spin: The angular momentum of atomic nuclei, arising from the angular momentum of protons and neutrons.

Observable: A physical quantity that can be measured experimentally, such as energy, position, momentum, and angular momentum.

Pair Production: The creation of a particle and its antiparticle from a photon. Examples include the creation of an electron and a positron, a muon and an antimuon, a proton and an antiproton, and a neutron and an antineutron.

Pauli Exclusion Principle: The principle stating that two fermions cannot simultaneously occupy the same quantum state; no two fermions can have the same set of quantum numbers within a quantum system.

Period: The time it takes for a wave to complete one full oscillation. The variable T represents period, and the unit is seconds. Frequency and period are related by the formula $T = 1/f$.

Photon: A fundamental particle of light, also known as a quantum of electromagnetic radiation. Photons have no mass, travel at the speed of light in a vacuum, and carry energy proportional to their frequency. The energy E of a photon is given by the equation $E = hf$, where h is Planck's constant and f is the frequency of the light. Photons exhibit both wave-like and particle-like properties, a phenomenon known as wave-particle duality.

Physics: Physics is a branch of science that seeks to understand the universe. It explores physical concepts and often explains them using the language of mathematics.

Planck's constant: A fundamental constant in quantum mechanics, with a value of $h = 6.626 \times 10^{-34}$ Js. The reduced Planck's constant is $\hbar = \frac{h}{2\pi} = 1.054 \times 10^{-34}$ Js. Planck's constant has units of angular momentum.

Probe beam: One of two laser beams used in saturated absorption spectroscopy. The probe beam has less power than the pump beam. In saturated absorption spectroscopy, the transmission of the probe beam is measured.

Proton: A particle found inside the nucleus of an atom. A proton has a positive charge of $+1.602 \times 10^{-19}$C, a mass of 1.672×10^{-27} kg, and a radius of 0.84×10^{-15}m $= 0.84$fm. A proton is composed of three quarks: two up quarks and one down quark. A free proton is stable and does not undergo radioactive decay.

Proton number: The total number of protons in an atomic nucleus, represented by the variable Z.

Pump beam: One of two laser beams used in saturated absorption spectroscopy. The pump beam has more power than the probe beam.

Quantum mechanics: A fundamental theory of physics that describes the behavior of atoms, subatomic particles, and molecules.

Quantum number: An integer (positive, negative, or 0) or half-integer used to represent a state of an observable, emphasizing the discrete nature of quantum mechanics. Quantum numbers have no units.

Quarks: Elementary particles that experience all fundamental forces, including the strong force. Each quark has an antimatter counterpart. Particles made from quarks, including protons and neutrons, are called hadrons.

Reduced Planck's constant (h-bar): Planck's constant divided by 2π: $\hbar = \frac{h}{2\pi} = 1.054 \times 10^{-34}$ J s.

Refraction: The redirection of a wave as it passes from one medium to another (e.g., air to water).

Resonance: When an atom is excited by a photon from one state to another, it "goes through resonance." This is similar to playing the trumpet, where blowing correctly into the trumpet excites a standing wave in the pipe to create a note. Similarly, when an atom is excited, the electron transitions from one standing wave mode to another. Atomic physicists use "excitation" and "resonance" interchangeably.

Resonance frequency: The frequency at which a laser excites an atom from one state to another. The variable f_r represents resonance frequency.

Saturated absorption plot: A plot representing the difference in the transmission of the probe beam with the pump beam on versus off. This technique removes Doppler broadening, leaving only the spectral features that would exist if the atoms were stationary.

Saturated absorption spectroscopy: A spectroscopic technique using a probe beam and a pump beam to produce spectral features from atoms with $v_{\parallel} = 0$.

Saturation intensity: The intensity of light required for an on-resonance laser to have 25% of atoms in a sample in the excited state at any given time. The variable I_s represents the saturation intensity.

Saturation parameter: The ratio of the laser intensity to the saturation intensity. The variable s represents the saturation parameter.

Scattering rate: The number of photons per second an atom absorbs and re-emits. The variable for scattering rate is $r_\gamma(\delta, s)$ when using linear frequency units and $r_\Gamma(\Delta, s)$ when using angular frequency units.

Schrödinger equation: The fundamental equation in quantum mechanics that describes how the quantum state of a physical system evolves over time. Using this equation, we can determine the wavefunctions and corresponding energy levels of the system.

Spectral component: "White" light is composed of many different wavelengths of light. Even light from the sun, which appears yellow, consists of various wavelengths. A single wavelength within a broader spectrum of light is called a spectral component.

Spectral feature: The lineshape on an absorption or transmission plot arising from the emission or absorption of a photon with energy corresponding to the difference between initial and final states of a transition.

Spectrometer: An instrument that separates and measures the amount of each spectral component.

Spectroscopy: A subfield of atomic physics that studies atoms through their interactions with light.

Standard Model of Particle Physics: A comprehensive theory of particle physics that describes the fundamental particles and their interactions, extending beyond quantum mechanics. It is one of the most successful theories to date.

Superposition: A concept in mathematics and physics stating that if multiple effects act on a system, the overall effect is the sum of the individual effects.

Torque: A measure of the effectiveness of a force in causing an object to rotate. The variable that represents torque is the lowercase Greek letter tau, τ.

Transmission plot: A plot showing the percentage or fraction of photons that pass through a vapor cell as a function of laser frequency.

$v_\parallel$: The velocity component of an atom in the direction of the laser beam. In the Doppler shift formula, $v_\parallel$ is negative if the atom is traveling towards the laser and positive if it is traveling away from the laser.

Virtual particles: Temporary fluctuations in quantum fields that arise from the energy-time uncertainty principle. They can mediate forces, like virtual photons in electromagnetic interactions, or appear as vacuum fluctuations, but they are not directly observable.

Wave interference: A phenomenon occurring when two or more waves overlap, resulting in a wave with greater (constructive interference) or lower (destructive interference) amplitude. Interference effects can be observed with all types of waves, including light waves, water waves, and electron waves.

Waist of a laser: The half-width of the intensity profile where the intensity is 13.5% of the maximum intensity, as shown in Fig. 3.7. The variable w represents waist.

Wavelength: The distance between two consecutive "like" points on a wave, such as two adjacent wave peaks. The variable that represents wavelength is the lowercase Greek letter lambda, λ.

Wave-particle duality: The concept that particles, like electrons and photons, exhibit both wave-like and particle-like behavior depending on the experiment, meaning they can act as waves in some situations and as particles in others.

Index

A

Absorption plot, 44, 47, 72
Absorption spectroscopy, 34
A general rule of nature, 164, 205, 213
Angular momentum, 137–139
    adding quantum mechanical angular
        momentum, 145–150, 177–181
    orbitals, 172
Angular units, 47
Antimatter, 229–231, 238
Atomic physics, 4

B

Basis sets, 121, 142
Beyond the standard model, 240
    baryon asymmetry, 240
    dark matter and dark energy, 241
    gravity, 240
    neutrino mass, 241
Blackbody radiation, 30–34
    the ultraviolet catastrophe, 31
Bosons, 172–173
Bra-ket notation, 58, 117, 152, 172, 180,
    216

C

Center of gravity, 58, 179, 185
Coulomb force/interaction, 160, 209
Crossover-free spectroscopy, 103–106
Crossovers, 90–97, 189
    V crossovers, 91, 189
    X crossovers, 94
    Λ crossovers, 94

D

Degenerate/degeneracy, 169–170

D (continued)

Detuning, 51
Diffraction, 26–27
Dispersive element, 26
Doppler broadening, 76
Doppler effect, 65
Doppler profile, 76, 87
Doppler shift, 70, 87
    astronomy, 79
Doppler width, 76, 77
Double-slit experiment, 8–13

E

Electromagnetic force/interaction, 166, 216
Electron, 5, 138, 238
Electron configuration, 164
Electron shells and subshells, 161–165
Equipartition Theorem, 77–78
Excited state fraction, 55

F

Fermions, 172–173
Feynman diagrams, 234–237, 239
Fine structure, 167–169
Fluorescence, 44
Force, 136
    Newton's 2nd Law, 136
Force carriers, 234, 238
Frequency, 10

G

Gluons, 234, 238
Graviton, 238

H

Higgs boson, 238

Hyperfine structure, 178–184
    amplitudes, 189
    electric quadrupole hyperfine constant,
        183
    hyperfine splitting, 58, 167
    magnetic dipole hyperfine constant, 183
    splitting formula, 183, 187, 191
    Wigner 6-j symbol, 189

I
Intensity, 50
Ion, 5, 174
Ionization threshold, 159
Isotope shift, 198–202
    field shift, 200
    normal mass shift, 200
    specific mass shift, 200

K
Kinetic energy, 77

L
Leptons, 238
Lifetime, 49, 228

M
Madeling energy ordering rule, 161
Mass number, 198
Maxwell-Boltzmann velocity distribution,
    73–75
Molecules, 5
Momentum, 114, 136
Muon, 238

N
Natural linewidth, 44, 47, 228
Neutrino, 203, 238
Neutron, 5, 138, 207, 234
Nuclear notation, 198
Nuclear shells, 217–219
    magic numbers, 217
Nucleon, 198

O
Observables, 114
    compatible/incompatible observables, 120,
        141, 151–153, 180
    the measurement game, 132

P
Pair production, 229
Pauli Exclusion Principle, 216
Pauli exclusion principle, 172
Period, 10
Photon, 17, 151, 238
Planck's constant, 19
    Reduced Planck's constant, 48, 128, 137
Polarization, 20, 151
    halfwave plate, 21
    polarizing beam splitter, 21
Positron, 204, 214, 230
    Positron Emission Tomography, 214, 220
Power, 20
Power broadening, 56–57
Prism, 26
Probe beam, 86
Proton, 5, 138, 207, 234
Pump beam, 86

Q
Quantum field theory, 232–237
    quantum chromodynamics, 236
    quantum electrodynamics, 236
Quantum harmonic oscillator, 127–128
Quantum mechanics, 7, 114
    Copenhagen interpretation, 17, 114
    limitations, 226
Quantum numbers, 137, 162, 178
    nuclear spin, 150, 177
    nuclear spin: z-component, 150
    orbital angular momentum, 137, 139,
        148–150
    orbital angular momentum: z-component,
        139, 148–150
    principal quantum number, 116, 162
    spin, 148–150
    spin: z-component, 148–150
    total electronic angular momentum,
        148–150
    total electronic angular momentum:
        z-component, 148–150
Quantum states, 116–119
    energy states, 15, 116, 128, 158
    position states, 117
Quarks, 205, 226, 239

R
Radioactive decay, 203–207, 226
    α decay, 204
    β^- decay, 203, 213, 239
    β^+ decay, 204, 214

carbon dating, 207
electron capture, 204, 215
half-life, 206
neutron emission, 203
proton emission, 203
spontaneous fission, 204
Reflection grating, 27
Refraction, 26
Resonance, 19, 42
Resonance frequency, 20, 42, 185

S
Saturated absorption plot, 90, 103, 186
Saturated absorption spectroscopy, 85–90
Saturation intensity, 54
Saturation parameter, 53
Scalar field, 231
Scattering rate, 51–54
Schrödinger equation, 116, 231
Scientific method, 6
Spectral component, 26
Spectral feature, 45
Spectral radiant exitance, 32
Spectrometer, 26
Spectroscopy, 5, 17–18
Spectrum, 26, 35
Speed of light, 19
Standard model of particle physics, 7, 237–240
Stefan–Boltzmann Law, 33
Strong nuclear force, 209–211, 234, 238
Superposition, 121–126

T
Table of isotopes, 207–208
Tau particle, 238

Term symbol, 166
The most important equation in all of science, 22, 131, 241
The rule, 59, 90
Torque, 136, 165
Transmission grating, 27
Transmission plot, 44, 47, 72, 102

U
Uncertainty principle, 121, 125, 128–130, 227–229
    Heisenburg's uncertainty principle (position and momentum), 129, 227
    Uncertainty principle (energy and time), 228

V
v-parallel ($v_\parallel$), 70, 85, 94
Vacuum fluctuations, 233–234
Vector field, 231
Virtual particles, 234

W
Waist of a laser, 50
W and Z bosons, 238, 239
Wavefunction, 116
Wave interference, 9, 10
    constructive interference, 10
    destructive interference, 11
Wavelength, 10
Weak nuclear force, 212–215, 238
Wien's displacement law, 34